THE GCHQ PUZZLE BOOK

The GCHQ Puzzle Book

PENGUIN BOOKS

PENGUIN BOOKS

UK | USA | Canada | Ireland | Australia
India | New Zealand | South Africa

Penguin Books is part of the Penguin Random House group of companies
whose addresses can be found at global.penguinrandomhouse.com.

First published 2016
015

Text copyright © Crown Copyright, 2016

Set in Plantin
Designed by Couper Street Type Co.
Printed in Great Britain by Clays Ltd, St Ives plc

A CIP catalogue record for this book is available from the British Library

ISBN: 978–0–718–18554–1

www.greenpenguin.co.uk

MIX
Paper from
responsible sources
FSC
www.fsc.org   FSC® C018179

Penguin Random House is committed to a
sustainable future for our business, our readers
and our planet. This book is made from Forest
Stewardship Council® certified paper.

Heads Together is a campaign being spearheaded by the Duke and Duchess of Cambridge
and Prince Harry, which aims to change the conversation on mental health from fear and shame
to confidence and support. This campaign wants everyone to feel comfortable talking about their
mental health and able to support their friends and families through difficult times.

Contents

To the hard-working men and women
of GCHQ – past, present and future

KENSINGTON PALACE

I have always been immensely proud of my grandmother, Valerie Glassborow, who worked at Bletchley Park during the Second World War. She and her twin sister, Mary, served with thousands of other young women as part of the great Allied effort to break enemy codes. They hardly ever talked about their wartime service, but we now know just how important the men and women of Bletchley Park were, as they tackled some of the hardest problems facing the country.

In a new century, their successors at GCHQ continue this intellectual tradition. Like their Bletchley predecessors, they have become well known for valuing and understanding the importance of mental wellbeing. This is so important when dealing with such discretion and the pressure which comes with this.

William, Harry and I are very grateful that this book is supporting our Heads Together Campaign. I hope it will not only amuse and challenge readers, but help to promote an open discussion of mental health problems, which can affect anyone, regardless of age or background. Together, we are aiming to change the national conversation around mental health from stigma and fear to openness and understanding. Those who buy this book and support the Heads Together campaign will be playing a part in helping people get the important mental health care they deserve.

Catherine.

Foreword by Director GCHQ

For nearly one hundred years, the men and women of GCHQ, both civilian and military, have been solving problems. They have done so in the pursuit of our mission to keep the United Kingdom safe. They have put their talents at the service of the nation's security through some of the darkest times in our history: the struggle against Nazism during World War II, when our staff at Bletchley Park played such a key role, and the battle to preserve the free world from totalitarian communism during the Cold War. Today, in this digital age, the same skills are being applied to fighting terrorism, countering cyber attacks and bringing the most dangerous criminals to justice.

Technology is at the heart of our work at GCHQ. Alan Turing created a novel electromechanical device to help break the Enigma code, and Tommy Flowers, our greatest engineer, built the world's first digital computer to help break the Tunny code. Our great mathematicians, James Ellis, Clifford Cocks and Malcolm Williamson, made the historic breakthrough which enabled the encryption upon which all of us who use the internet rely for our security and safety everyday: Public-key cryptography.

But at the heart of the mathematical, engineering and linguistic successes for which GCHQ is famous are individuals with skills, talent and commitment. As well as puzzling for the national good, many of them also do it for fun. They have created this book to share their hobby with you. I hope it will entertain you and perhaps inspire some younger readers to become our problem solvers in the future.

At GCHQ we are excited by the dazzling possibilities for human development offered by the internet, but we also know that this technology can be abused by a minority who seek to do harm. If we are to keep meeting these threats, we need creative, expert and dedicated staff. GCHQ has a proud history of valuing and supporting individuals who think differently; without them we would be of little value to the country. Not all are geniuses or brilliant mathematicians or famous names, but each is valued for his or her contribution to our mission.

Over the decades, we have come to understand the pressures on mental health which affect so many, regardless of age or background. I am therefore delighted that the proceeds from this book will go to the Heads Together group of mental health charities, which Their Royal Highnesses the Duke and Duchess of Cambridge and Prince Harry have done so much to encourage.

I want to thank all those GCHQ staff who have given their own time to create these puzzles over the years, and particularly those involved in creating this magnificent collection. I cannot name them, for obvious reasons, but I know how much they have put into this project.

Thank you for buying this book and supporting these charities; all of us at GCHQ hope you enjoy it.

Robert Hannigan
Director GCHQ
Cheltenham, September 2016

Solving Puzzles for Fun at GCHQ

Solving serious puzzles is at the heart of what GCHQ does, whether analytical, technical or cryptographic. It is not surprising then that many of us employed in GCHQ also enjoy setting and solving puzzles for fun.

For many years it was codes which formed the bulk of such puzzles, and they were also used in our recruitment literature. A less purely cryptographic puzzle/quiz began in the 1980s and started a tradition of puzzles for people to solve over the Christmas period. Typical questions were:

> M, N, B, V, C, X, ?
>
> BAGG is to William the Conqueror as BEJC is to whom?
>
> Which of GLITHGINRYO, UDLOTWIN, TIPSYCHATRY, CHASTIPLAW might complete the sentence: THRIPGUH is to ALMOOW as GUMP is to?

(Answers to questions in this chapter are given at the end of the chapter.)

After a short gap in the 1990s this was reborn as the Christmas Puzzle Quiz (CPQ), which continues to this day – and many of the puzzles in the book come from this source. The CPQ has always been a mix of general knowledge and puzzle solving, often including sets of questions connected to Cheltenham itself, and some questions where there doesn't even appear to be a question. For example:

> 42°15'N 72°15'W, 53°52'N 44°50'E, 37°49'N 85°29'W, 39°37'N 75°56'W, 40°57'N 40°17'E, 51°54'N 02°04'W?

The CPQ has raised over £3,500 for various charities over the years. Like all of the puzzles and events I will describe here, the puzzle creation and compilation is done outside the office.

An altogether tougher puzzle appeared for the first time in 1993, also to be solved over the Christmas period. This evolved into the Kristmas Kwiz, and concentrated more on letter, word, number and code puzzles. As time has moved on, this has become more and more popular within GCHQ as well as with our international partners, with entrants from seven countries now taking part. Again many of the puzzles in this book were originally set as part of a Kristmas Kwiz. A typical Kwiz kwestion might be:

> What word might finish the following sequence:
> APE, VOL, TRAIT, REN, EURO, ON, LID, ENTICE, ROD?

Towards the end of this book you will find a complete Kristmas Kwiz challenge, giving you the opportunity to match yourself against the puzzle setters and solvers of GCHQ. Read the introduction to the challenge on page 121 to see how it works, and where and when you can find the answers.

Over the next few years a number of other puzzle events were created, not least to ensure that people's brains could be kept active through the whole year and not just at Christmas. The first was the Treasure Hunt in 1997. This means that one Sunday a year about fifty of us descend on some town in the Cotswolds and walk around with clipboards solving puzzles that lead to a Treasure. The idea of a Treasure Hunt is not unique to GCHQ, of course, but the nature of some of the questions perhaps is. For example:

> What is $2m(4a-3p)$?

Calculating this equation gave the answer 8am–6pm, which referred to local parking restrictions. Not all questions are like this though. For example: 'What is the most fashionable place around here?', to which the answer was 'This postbox' – as a label on the nearby postbox read: THIS POSTBOX HAS THE LATEST COLLECTION IN THE AREA!

In 1999 an event began which I believe is unique to GCHQ. The style is that of a pub quiz, in that teams of four spend an evening answering rounds of questions, most of which are read out. However, none of the questions are general knowledge, they are all puzzles. There might be rounds of word puzzles or number puzzles, but over the years there have also been memory rounds and 'physical rounds' – such as colouring a toilet roll to demonstrate a mapping problem, balancing eggs on towers, or deriving a code from tasting jellybeans.

To give a flavour of this event here are a couple of questions:

> Yesterday I looked out of the window and saw something. I then glanced at my watch and noticed two things:
>
> * it was 3 minutes past 8
>
> * what I had just seen through the window was written on my watch.
>
> What had I seen?

> Last night I wrote down the numbers from 1 to 99. I then rearranged them into alphabetical order. Which number didn't move?

In 2004 the idea arose of running a puzzle over the Easter period, and the Easter Teaser was born. If you noted the anagram then you might appreciate that the 2005 version was entitled 'Easter Teaser – a Reset'! This has taken a variety of forms over the years, one of which ended up appearing as a competition in *The Times* in connection with the 2011 Cheltenham Science Festival.

An early Easter Teaser question was:

> Mr Green likes bottles of champagne, minstrels and fallen women. What's his favourite game?

Most recently the annual Puzzle Hunt, which is similar to other existing puzzle hunts, has been added to the calendar since 2014. This consists of a set of pictorial puzzles which have to be completed by teams in one evening. Although the answer to each puzzle is a word or phrase, none of the puzzles actually have questions. The aim is to work out what to do from the picture itself, sometimes using a hint in the puzzle title. At the end of the evening the answers you have found fit into a final meta puzzle, which itself has no question. As with the Kristmas Kwiz mentioned earlier you have an opportunity to match yourself against us – the first colour section contains a complete Puzzle Hunt, including a Meta Puzzle. Why not see how you and your friends get on over an evening?

Although the puzzles and events described above are enjoyed by GCHQ staff and their families, with some also shared with colleagues in partner agencies abroad, there have been some special occasions when we have created puzzles for people outside GCHQ. One such was the centenary of Alan Turing's birth in 2012, which was commemorated by an exhibition of his life and work in the Science Museum. As part of this event a few of us produced a set of puzzles aimed at children. For example:

> What thirteen-letter placename is hidden in this sentence?
>
> You need the ability to receive clues e.g. noticing exactly 13 examples of one vowel in a sentence – the letters just before every appearance of one of a, e, i, o or u may inspire your brain to spot the trickiness.

In 2013 we were asked by the magazine *Physics World* to produce some physics-themed puzzles to celebrate their 25th anniversary. Five appeared around the anniversary itself, followed by another that Christmas, and one again the following Christmas. This was the first time we'd really had an opportunity to see how popular these sorts of puzzles were beyond our own community – over 5,000 people solved the first one, and we were delighted to see the large number of positive comments posted on the Physics World website.

The first puzzle, repeated here by kind permission of *Physics World*, was:

> There is a word missing from the following. What is it? (Give the answer in its encrypted form.)
>
> TNVERI SMH EG ZSMRNPMUD: M SLRN PYMP
> VERRNVPT M ZSMRNP PE PYN TQR THNNZT EQP
> NXQMS MUNMT LR NXQMS PLKNT

The full set of puzzles can be found at: www.physicsworld.com/puzzle.

More recently still, in 2015, the Director of GCHQ asked some of us to create a puzzle which he could include in his official Christmas card. This needed to be something which would appeal to as many people as possible, but which ideally would not be completely solvable by anyone. The resulting challenge took the form of a Nonogram (or Hanjie) in the actual card itself (reproduced in the first colour section). Correctly completing this led entrants to a web address, and a set of multiple choice puzzles. Solving these led in turn to a set of word puzzles, then number puzzles, and finally a large number of different sorts of puzzles, some hidden, and some with more than one answer.

It is still possible for you to attempt the Christmas Card Puzzle, either by solving the Hanjie or by going to: https://www.gchq.gov.uk/puzz.

The response to the Christmas Card Puzzle was breathtaking, and global. More than half a million people tried to solve the first stage, and more than 10,000 of these reached the final stage. Those trying came from all over the world, and 550 people submitted answers. No one completely solved the puzzle, although six people got very close – and three of these were chosen to win prizes based on the quality of their reasoning. Those who enjoyed the Christmas Card Puzzle were encouraged to donate to the NSPCC. Similarly, all the profits from this book will be donated to Heads Together – a grouping of eight mental health charities which together are tackling stigma, raising awareness, and providing vital help for people with mental health challenges.

It was the unprecedented level of interest in the Christmas Card Puzzle which led to the idea for this book, which contains a whole range of different puzzles set by staff who work, or have worked, at GCHQ.

We hope you enjoy it.

Answers to questions in this chapter:

M, N, B, V, C, X, ?

Answer: **Z**. The sequence is the bottom row of a typewriter keyboard in reverse.

BAGG is to William the Conqueror as BEJC is to whom?

Answer: **1492. Christopher Columbus**. BAGG represents 1066, with each letter representing its place in the alphabet minus one, i.e. A=0, B=1, C=2, etc. BEJC is therefore 1492.

Which of GLITHGINRYO, UDLOTWIN, TIPSYCHATRY, CHASTIPLAW might complete the sentence: THRIPGUH is to ALMOOW as GUMP is to?

Answer: **UDLOTWIN (uDlOtWiN)**. Reading every second letter in the groups in the sentence gives HIGH is to LOW as UP is to DOWN.

42°15'N 72°15'W, 53°52'N 44°50'E, 37°49'N 85°29'W, 39°37'N 75°56'W, 40°57'N 40°17'E, 51°54'N 02°04'W?

Answer: **52°12'N 1°41'W**. The coordinates in the question are the latitudes/longitudes of: Ware, Issa, Bardstown, North-East, Of, Cheltenham. Read as the question: Where is a Bard's town north-east of Cheltenham, the answer is Stratford-on-Avon – and the answer is its latitude/longitude.

What word might finish the following sequence:

APE, VOL, TRAIT, REN, EURO, ON, LID, ENTICE, ROD?

Answer: **TART**. From the sequence you can form new words by adding the same letter at the start and end, and these words spell out Los Angeles. In fact any word from which you can form a new word by adding an S at the start and end is a correct answer.

LapeL, OvolO, StraitS, ArenA, NeuroN, GonG, ElidE, LenticeL, ErodE, StartS

Yesterday I looked out of the window and saw something. I then glanced at my watch and noticed two things:
* it was 3 minutes past 8
* what I had just seen through the window was written on my watch. What had I seen?

Answer: **MOON**. My watch said ᴇᴏ:ᴏᴇ, which when read sideways spells MOON.

Last night I wrote down the numbers from 1 to 99. I then rearranged them into alphabetical order. Which number didn't move?

Answer: **sixty-nine**.

Mr Green likes bottles of champagne, minstrels and fallen women. What's his favourite game?

Answer: **Othello**. The question refers to English translations of the Italian: Giuseppe Verdi (Mr Green) composed *Otello*, *Nabucco* (Nebuchadnezzar), *Il Trovatore* (The Troubadour) and *La Traviata* (The Fallen Woman).

What thirteen-letter placename is hidden in this sentence?

You need the ability to receive clues, e.g. noticing exactly 13 examples of one vowel in a sentence – the letters just before every appearance of one of a, e, i, o or u may inspire your brain to spot the trickiness.

Answer: **Bletchley Park**. As described in the sentence, one of the vowels, i, appears exactly 13 times. The letters which appear immediately before each i spell Bletchley Park.

There is a word missing from the following. What is it? (Give the answer in its encrypted form.)

TNVERI SMH EG ZSMRNPMUD: M SLRN PYMP VERRNVPT M ZSMRNP PE PYN TQR THNNZT EQP NXQMS MUNMT LR NXQMS PLKNT

Answer: **KEPLER**. The text reads: Second Law of planetary: A line that connects a planet to the sun sweeps out equal areas in equal times. The missing word is MOTION, and this encodes to Kepler, whose law it is. The cipher alphabet in full is:

Plain: ABCDEFGHIJKLMNOPQRSTUVWXYZ
Cipher: MOVINGBYLAWSKREZXUTPQJHFDC

which begins appropriately: 'moving by laws'.

Introduction to the Puzzles

The puzzles in this book range in difficulty but we hope there is something for everyone.

Roughly speaking the puzzles tend to get harder as you progress through the book, but this isn't precise – so if you have solved an early puzzle it may still have been a hard one. In multi-part questions the later parts tend to be harder than the earlier parts.

For the first section, 'Starter Puzzles', we have provided clues for each question, which can be found on pages 151–3.

Both these questions and the later ones will become easier once you are attuned to the mindset of the setters – there are several repeated themes and styles which occur throughout. Because of this we have designed a way in which you can get help without cheating – by using the 'Hints' section (on pages 137–43). In this we explain how to approach some questions and also give various themes and topics which you can consider if you are struggling with a puzzle. These are all presented as general hints, not specific to any particular question. Hence by looking at the Hints, you aren't cheating – so you can still give yourself full credit when you solve the puzzle you were stuck on.

Some styles of puzzles (called 'Where' and 'Which') appear several times throughout the book, and how these work is described in detail on pages 145–9.

As an additional aid, particularly for those of you trying to answer questions whilst on a plane, in a jungle, or halfway down the M4 – in other words where you may have no signal to access the internet – there is an Appendix which contains some useful lists of information and which, like the Hints section, may help you if you are struggling.

As many of the questions, or kwestions, have come from the Christmas Puzzle Quiz and the Kristmas Kwiz, you will find a few Christmas references throughout, and also mention of the kwiz koffee kup. This is a notoriously unstable drinking vessel which tends to get knocked over and spill koffee over crucial parts of certain kwestions!

As well as the puzzles in the main sections, there is the Kristmas Kwiz Challenge on pages 121–36, and also a Competition, the first part of which can be found in the first colour section. Other than for these, the answers to puzzles are in the answer section at the back. And of course there may be more questions in the book than those which are immediately obvious . . .

We hope you enjoy the puzzles in this book, whether you start from the beginning and work through, or just dip in and out of the pages.

Section 1: Starter Puzzles

Individual clues to the puzzles in this section can be found on pages 151–3.

1. **The early bird**
    What can be African, Emperor, Little or Macaroni?

2. **Replace the stars**
    What is missing?
    Picture Menace, Khan Clones, * *, Home Hope, Frontier Back

3. **Calendar foods**
    If we eat Fish on St David's Day, Crab on US Independence Day and Goat on Christmas Day, what do we eat on Halloween?

4. **Don't be hasty**
    (a) What is a 100th of a dollar called?
    (b) Which county is known as the garden of England?
    (c) What is the period of fasting that precedes Easter called?
    (d) What has types including Bell, Frame and Oxygen?

    What links these answers – and which author created these creatures?

5.   Limerick

This limerick has a different 5-letter word missing from each line. What are the missing words? The words form a 5×5 square.

A man from the East, name of - - - - - ,
Would - - - - - give people red roses,
But he had an odd - - - - - ,
A strange - - - - - glare,
Which is why they would - - - - - , one supposes.

6.   Where next?

What follows Vienna, Brussels, Prague, Copenhagen, Tallinn, Helsinki, Athens, ?

7.   Aha

(a) Which canoe, created by the Inuit, is propelled by a double-bladed paddle?

(b) Which note is equivalent to two crotchets?

(c) What is another name for midday?

(d) According to the Bible, what was the name of the first woman?

(e) What do you call a female sheep?

How would you 'refer' to these answers?

8.   Concealed animal

Here is a list of animals. But which other animal is hiding?

RACCOON, PARROT, LEMMING, KOOKABURRA, AARDVARK, LINNET, OTTER

9. **Chromatic separation**

The following list has been divided into three sets, in three different ways, according to simple rules. What is the rule in each case?

a. BLACK b. BLUE   c. BROWN   d. GREEN   e. GREY
f. MAUVE g. ORANGE   h. RED   i. WHITE   j. YELLOW

i. acdfi / beghj
ii. bdghj / acefi
iii. acdei / bfghj

10. **What is polite?**

If 1000110 is MINGLE and 1000110001100 is APE, what string of digits is POLITE?

11. **When, how, what?**

(a) In which year will the Olympics be held in Tokyo for the second time?

(b) How is perfect vision described?

(c) What is the shortest form of international cricket?

12. **Odd words out**

Find the odd word out in each of the following lists:

(a) ARM, ELECTRIC, LEAGUE, PUSH, ROCKING, VICE, WHEEL

(b) ART, DOG, FREE, HALFWAY, PUBLIC, PUBLISHING, ROSE

(c) DANCE, GLASS, GROUND, OCEAN, SECOND, SHOP, TRADING

13. **A round of drinks**

What could follow Mojito, Eggnog, Riesling, Lemonade, Ouzo, ?

14. Cheek

(a) Which Carla Lane sitcom concerned the Boswell family and their attempts to keep solvent?

(b) What was the surname of the British Prime Minister from 2007 to 2010?

(c) The institution which produces British coins is called the Royal . . . what?

(d) Which fruit has varieties beef, cherry and plum, and is normally served as a vegetable?

(e) Which county was merged with Herefordshire in 1974?

What connects these answers?

15. Letter sequence I

What is the next entry in each of the following sequences:

(a) M, V, E, M, J, S, U, ?

(b) E, Z, D, V, F, S, S, A, N, Z, ?

16. Missing and not missing

Eric, Graham, John, Michael, Terry. Which name is both missing and not missing?

17. Somewhere

(a) What does UV stand for?

(b) Which Duke Ellington jazz standard has also been recorded by Annie Lennox and Nina Simone?

(c) Which boy band represented the UK in the 2011 Eurovision Song Contest?

(d) Which piece of music is believed to have been written by Henry VIII?

(e) In which US national park is Old Faithful?

(f) What do the bells of St Clements say?

(g) Which horse won the Grand National in 1973, 1974 and 1977?

What might you find at the end?

18. Friends in other countries

Gary has friends in countries all around the world including Argentina, Denmark, Russia, Sudan and the United States. Which country is Gary in?

19. Red and Black

If Broken/Purple and Rough/Baseball are red, and Sam/Garden and Golf/Fight are black, how many are in each?

20. Who?

If India=1, Victor=5, X-Ray=10, Lima=50, Charlie=100, Delta=500, who=1000?

21. Composing a sequence

What is the final entry in this sequence?

Brahms's 1st, Elgar's 1st, Mahler's 5th, Schubert's 8th, Schumann's 3rd, Borodin's 2nd, Shostakovitch's 9th, Bruckner's 7th, Beethoven's ?

22. A question to put you to sleep

Find a whole number which when spelled out has its letters in alphabetical order.

23. Link

What links: D, X, 4th, Y

24. Pairs

The answers to these questions come in pairs where each pair differs only in their first letter. Work out the pairs. The pronunciation may not always be the same.

1. One half of a titular TV sitcom couple.
2. British budget clothing and homeware retailer.
3. Type of hat.
4. One of the Channel Islands.
5. River that passes through northern English city.
6. TV space alien.
7. Capital of country with a coastline on the Persian Gulf.
8. Surname of an England football captain.
9. An Irish county.
10. An English city that is a county town.
11. One of the official languages of the fourth largest country in Europe.
12. District of South-East London.

25. Who looked out?

SQUAW, SCAMP, ASTER, BASIN, HALLO, CLOTH, RATIO, PARSE, HEARS, CARES, GRAVE, CHORE, ASSES

26. Missing word

What word is missing: AAGKN, ABBIRT, EEEORY, EGGIRT, EGILPT, LOW, OOR

27. Numerical sequences

(a) A, C, F, J, O, ?
(b) A, B, D, H, ?
(c) B, C, E, G, K, M, Q, S, ?

28. Replace the ?

There are at least 20 children in Fred's class. 19 of them learn German and 4 learn French. 8 of them play football, 12 play hockey and 4 play squash. Fred lives at number 18 Bunyard Street in a house with a red door. It takes Fred ? hours to mow a 1 acre field.

29. Odd one out

Which is the odd one out: FIRM, HELM, SOAK, WASH

30. Fill in the blanks

Fill in the blanks: ? was a rolling stone, ? beach, ? California, ? in the jar, ? and ?

31. Olympic sports

Which Olympic sport completes the following sequences? (Some may have more than one possible answer):

(a) BASKETBALL, WATER POLO, FOOTBALL, BEACH VOLLEYBALL, VOLLEYBALL, HANDBALL, FIELD HOCKEY, TENNIS, GOLF, ?

(b) SYNCHRONIZED DIVING, MEN'S TEAM SPRINT, BASKETBALL, HANDBALL, ?

(c) ARCHERY, TRIATHLON, HANDBALL, LUGE, EQUESTRIANISM, TAEKWONDO, ICE-HOCKEY, CURLING, ?

32. Literature

Whose works include:

(a) OT, ACC, TOCS, ATOTC

(b) ASIS, TSOF, THOTB, TVOF

(c) S, VB, DAF, AHOD

33. Sums

Black × (Pink + Blue + Brown + Green + Yellow + Red) = ?

34. Which word?

The following list of words can be divided into two sets, of equal size, on the basis of another word:

AWE	BODY	DAY	FEAR	HAND
HOW	LONE	ONE	THING	TIRE
WHAT	WHERE	WHOLE	WIN	

What is the other word?

35. Stupid riddle

My first is in FIRST, but not in ONE.

My second is in SECOND, but not in TWO.

My third is in THIRD, but not in THREE.

My fourth is in FOURTH, but not in FOUR.

My fifth is in FIFTH, but not in FIVE.

Important note: The Whole describes the puzzle setter, because he forgot about the O in FOUR.

36. Explain

If 355 equates to 524, and 1235 to 2521, what does 850 equate to?

37. The magic words

What four words come before all the following: 1. PS, 2. COS, 3. POA, 4. GOF, 5. OOTP, 6. H-BP, 7. DH

38. Musical plant?

Musically, what kind of plant is usually seen in two forms:

Single – most often for C and S

Double – most often for B and O

39. How many people?

If 1 person can be considered LONELY, but 2 people are a NETWORK, how many do you need for them to become THREATENING?

40. Names

Apart from being names, what do the following have in common:

ALICE, CICELY, ELAINE, ESTHER, JANE, JASON, MARIAN, NORMA, RONALD

41. Shopping list

Tidying up the other day I came across an old shopping list that I wrote out for a friend who was poorly last Christmas. What was his name?

Pizza	Eggs	Rioja	Figs	Orange juice
Razor	Mixed nuts	Frying steak	Rusks	Emery boards
Quiche	Unsweetened squash	Expectorant	Nectarines	Cabbage
Yoghurt	Coffee	Oven chips	Underwear	Newspaper
TV guide				

42. Premier locations

If C=SB, CP=SP, E=GP, MU=OT and TH=WHL, what does L= ?

43. What links?

(a) What sort of plating replaced nickel electroplating in car manufacturing?

(b) The lead guitarist of U2 is known as The . . . what?

(c) What kind of performance is staged at La Scala?

(d) What word derives from a Swahili word meaning 'journey' but is now more commonly used to mean a wildlife-viewing trip?

How are these answers linked?

44. Next country

Which is the only country that can come next in this sequence:

Cyprus, Sweden, Morocco, ?

45. Missing body parts

Some body parts have been removed from the following words
(e.g. el*sis = elLIPsis):

hackn*d	childb*ing
m*alade	hor*le
bewit*g	de*ance
arc*elago	na*cond
batt*round	or*ra
astoni*g	ar*ent
pota*s	cri*ss
merc*ise	th*ss
freew*ing	p*er

46. Question time

(a) M, W, F, S, T, T, ?

(b) WR, SG, SR, ?

(c) ... , J, A, S, O, N, ?

(d) If 7 + 8 = 3 and 10 + 9 = 7, what is 11 + 2?

47. Metro access

(a) Red admiral

(b) Uncouth president

(c) Real bacon

(d) Synthetic cream

(e) Gilded rhombus

48. **What word could complete?**

What word could complete:

MORNING, NEAR, CHLORINE, TUXEDO, EXAMPLE, EFFECTS, CENTRE, COMPUTERY, AUSTRALIA, POSTSCRIPT, NAVY, BRITAIN, ?

49. **Game scores**

If Monopoly scores 15 and Cluedo scores 9, which board game scores 14?

50. **Where next?**

1 2 6
4 5 9
3 7 8

Where does 10 go and why?

51. **Can you?**

If you can dramatize: unwills, foliate, gumshoe, weakest, respect, scaredy

Then

(a) illustrate: comical, echelon, spangle, ocelots

(b) calculate: impiety, current, inferno, Sumatra

52. **Odd one out**

Which is the odd one out: Romeo, Two Zero, Brown, Force, Liverpool

53. **What's next?**

If 3=T, 4=S, 5=P, 6=H, 7=H, what is 8?

54. Go looking

Suggest a name for number 3:

1. ASA
2. IVY
3. ?
4. ARMAND
5. MEL
6. ARI
8. ART
10. CAT

55. Two words

The following 18 words fall into two sets of nine, according to two other words:

> DAM, DRONE, INLAND, INTER, KING, LACE, LADY, LICE, LINGER, NICKED, RENTAL, ROLE, SON, STIFF, STING, TINS, USED, WING

What are the two words?

56. Rooms

If Study ⟷ Kitchen, then Conservatory ⟷ ?

57. For the ears

What links the following, and where might you find them all together?

Ear, Champagne, Shoe, Bermuda, Ear – and Super

58. **Find N**

In the following, 'N' sometimes represents two words. What are they?

> N is W of S
>
> S is N of E
>
> E is E of N
>
> W is W of E

59. **Odd one out**

Which of the following words is the odd one out?

CHAT, COMMENT, DIRE, ELF, FORT, MANGER, PAIN, POUR

60. **Age-old question**

In 2011, members of the Riemann family were aged 11, 13, 41 and 47. Which year will be the next to be quite so special for them?

61. **Maze**

This is a maze. The start is the T in the 8th position on the top row. But which letter represents the 'centre' of the maze?

T	U	O	R	S	I	H	T	H	I
E	I	U	O	R	S	I	H	I	S
T	S	T	H	D	E	R	A	S	R
H	G	E	E	R	R	I	S	I	O
E	N	I	S	N	O	N	E	T	U
O	O	R	W	G	R	G	I	S	T
N	H	T	A	P	W	E	H	T	E
E	E	A	C	H	N	D	R	O	I
W	R	H	S	E	E	T	R	C	S
H	I	C	T	H	E	C	E	N	I

62. One way or another

Solve these one-dimensional anagrams: Salty Sherry, Rain on Hall, Alpine Yam, A Lazy Mink, I Insult Solomon

63. Who?

If Hart = 1, Ought = 2, Twee = 3, Bake = 4 and 6, then what number is Coy?

64. Arrange the pieces in alphabetical order

ARCH, FROGS, PLAN, WEE

65. Triangle

What are the top 3 letters in the triangle?

```
                ?

             ?     ?

          O    O    Y

        T    O    L    A

      I    E    F    F    L

    M    S    B    E    O    P

  I    F    U    C    H    D    E
```

66. Off the rails

What's the final member of this sequence: Central, Circle, District, Bakerloo, Piccadilly, Hammersmith and City, ?

67. Three sets

Arrange the following into 3 sets:

a. BORODIN   b. COLOSSUS   c. DIAMONDS   d. ENVY

e. GREEN   f. HAPPY   g. JACKSON (MICHAEL)   h. JACKSON (MILT)

i. KELVIN   j. LUKE   k. SMELL   l. SOLON

m. SPRING   n. SUNDAY   o. SUPERIOR   p. U

68. Spot the link

What links: a Golf course, Website navigation, Deodorant, Nintendo characters, and a Cat.

Section 2: The Puzzles

1. Where?

Where does RESURRECTION fit into the following list? *(Read left to right, top to bottom.)* See pages 145–8 for explanation.

PROGRESSION	COOT	MEAL	CHORAL	THUNDER
IDENTITY	SIMPLE	ROMANTIC	COPY	MANUSCRIPT
BLUE	CUCUMBER	BEECH	LORD	ENGLISH
FIDDLE	CONTENDERS	RUSH	GOLD	SURPRISE
KITE	COVENANT	AGENDA	AGE	PENCIL
CIRCLE	PAPER	ITALIAN	PRAGUE	LIGHTING
CYCLE	HILLS	WEEKEND	TENT	MOSAIC
BLONDE	CLASSICAL	BEETROOT	EXCHANGE	INHERITANCE
TRAGIC	SPRING	SCREEN	JUDGE	DRUM
SEA	GHOST			

2. What next?

What is the next word in this sequence:

UNSAID, RANDOM, SALUTED, DANEWEEDS, DRAUGHTS, AFRAID, ?

3.   Odd one out

Which is the odd one out?

(a) BEGGAR, BOXCAR, DELIVER, HARDWARE, LITIGATION, MOLECULE, SCOWL, SEVERAL, VOLATILE

(b) BAIT, BALL, BIRD, GUARD, JACK, LEG, LIST

(c) BODY, HOTEL, KNIFE, RELAXING, STATELY, TROUBLES, UNGRATEFUL, WRINKLE

(d) ARSENAL, CARAFE, CIPHER, COTTON, IDEA, MAGAZINE, MATTRESS, MONSOON, SOFA, SYRUP

(e) CASTLE, DEATH, HARBOUR, IDEA, LITTLE, RECREATION, TYPE, WRONG

4.   Lists I

Each of the following lists leads to a second list, which leads to a third list, which leads to a fourth list which is in alphabetical order. What are the fourth lists?

(a) GODS, VASE, GROWN, STAPLER, EARTHY, DIET, CANED, SERVE, SADDLER, SNAP, BLOTS, PLACER, PRIESTS

(b) BIRDWATCHER, BOMBING, BEEF STEW, DAGGER, BROAD-BRIMMED HAT, OPENNESS, DEAD END, OPEN PIE, GRIEVANCE OFFICIAL, DRUMBEAT, PAPER-FOLDING

5.   Please stand

In the following, what does the ? stand for?

UNITED KINGDOM	3435
AUSTRALIA	794
UNITED STATES OF AMERICA	3486
NEW ZEALAND	?
CANADA	16

6.   Odd one IN

(a) Which of the following words is the odd one in (i.e. not the odd one out!), and why?

PRIEST, PARSON, PADRE, CURATE, PEAHEN

(b) Which of these numbers is the odd one in, and why?

4, 66, 121, 484, 1936

7.   Identify the film

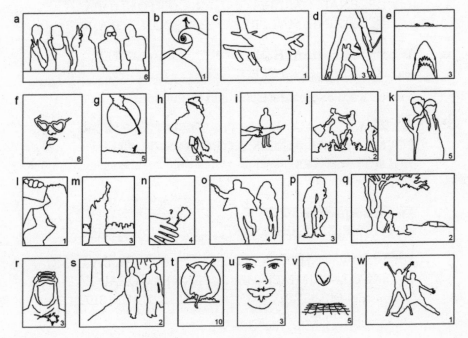

8.   Odd sequence out

Which sequence is the odd one out and why?

(4 8 11 13), (9 18 7 2), (16 13 1 19 8), (20 18 9 7 8), (4 12 7 15 9 8),
(7 1 15 20 8 19), (10 1 5 14 8 18), (13 1 16 7 11 17), (16 1 9 19 11 20),
(7 3 18 12 6 1 9)

9.   Properties I

The words to the left of the colon all have a (non-trivial) property that the word to the right of the colon doesn't have. What is that property?

(a) BLACKENED, DEFACING, FREIGHT, HIGHJACK, POLICEMAN, REQUEST, OVERTURNS: CERTAINLY

(b) ERROR, OUTER, PITY, QUIET, TORQUE, WRITER: FLASH

(c) BENZENE, BETWEEN, COLOURFUL, CRUX, FOIL, FRILL, GAD, JAMMY, KNEE, PAMPAS, VAMP, WHEEZE: GASEOUS

10.   What's in common?

What do the following words and phrase have in common? Suggest another word or phrase which could be in this list.

AGEISM, ALES, CAPE, EARTH, EMIRATES, EMIGRE, FOEMAN, LIMA, MALLS, NOTE, SOONTIME, THUMPING NOISE

11.   Word sequence I

What is the next word in the following sequences (in most cases there is not a unique solution, so any word which fits will be correct):

(a) CHIN, PLUM, MAGI, HEAR, FORT, BRIE, CLAN, CLOT, ?

(b) PIECE, LADDER, HASTE, EMOTION, BONY, LITTER, RASPING, ARROW, RISES, ?

(c) HERD, TALE, FOAL, CAGE, CRAM, RILE, TIER, ?

12.   A literary question

Richmal Crompton, Johanna Spyri, Lewis Carroll, Wilbert Awdry, Vladimir Nabokov, Alexander Solzhenitsyn, Arthur Conan Doyle, Rudyard Kipling, Benjamin Disraeli, Harriet Beecher Stowe, Henry Wadsworth Longfellow, Frederic Farrar, Joyce Lankester Brisley?

13.   Can't play it

What is the shortest word that can legally be played in a game of (British) Scrabble, but which can never appear on a Scrabble board?

14. Christmas songs

(a) O-LITOOFBEHOSTWESETHLI

(b) ECHEALITOUF-OLA-A-A-A-A-A-A-A-A-

(c) EYIDER--OW--RIEYIDER--AC--RT

(d) LEPHCHIRCK--HT

(e) LEEYLEEY----Y---

(f) TETETU--------NETE

(g) --MA----MA----------GI

(h) --N---IN----------

(i) ----L--------------LY

(j) -------------T

15. Find the link

What links?

(a) GARRISON, PORK, FIERCE, GIANT, HARES, TART, HOOKER

(b) BRA, CARE, MILLIONAIRE, NEEDLE, PRINCE, TIMELINE

(c) 10, 190, 2766, 57005, 11325150, 14613198, 16435934

(d) BUTCHERS, DIVER, HANDY, HARLEM, LARGE

16. 13 Pairs

Add one word to the 24 listed below to give 13 (yes, 13) pairs.

AFT, AFTERS, ALOFT, FELT, FILE, FUSTIER, GLARE, GROWN, HATED, HATRED, HINT, ISLE, LAYER, LOWERS, MALLS, PEACH, READ, SCENTED, SERVE, SKIN, SPORE, STANCE, TALE, THRUST

17. Identify the following

(a) Author of 21f.

(b) Weapon that fires a shot of diameter approximately 4.57mm.

(c) Poet.

(d) Chicago serial killer.

(e) Opening line of a Leiber and Stoller lyric.

(f) TV writer/producer and film director, born in 1966.

(g) Rapper, born in 1968.

(h) 16.5mm track.

(i) A tourist attraction in Bristol since 1970.

(j) Rock band noted for their beards.

18. Where?

Where does WHITE fit in the following list? *(Read left to right, top to bottom.)* See pages 145–8 for explanation.

UP	SALLY	PENNSYLVANIA	SHY	HUDSON
PUSS	MIDDLE	ENTER	GLENN	PIECE
GERMAN	FERTILE	THREE	HAND	BIRTH
SURGEON	STUPID	VITAMIN	CHRISTMAS	BOWL
CLOCK	MOSES	SURLY	CELLBLOCK	CHEERFUL
MAKE	BONE	BLUE	SPECIAL	PENNY
FALL	SKY	MAGGIE	EARTH	LET
STING	ABBEY	FINAL	WORD	LIVE
SLEDGE	BOARD	TIRED	SNUFFLY	LIE
SESAME	MODEL	RIVER	FOOTBALL	JUST
SAM	HENRY	DEATH	HIGH	MALCOLM

19. What's the missing word

(i) NEQQA WF Z IFCKDKAQ

(ii) LKTUL NJYX YWHCSUXE

(iii) VFJQUA BJ D CKDNQIN

(iv) DCGYCZ XM G PXBF-DCGP ODXYCF

(v) WKPIY OP T POWNKG

(vi) NHSD WS H DWL-YHNF

(vii) FPHPWMG PL KBR _ _ _ _ _ _ _

20. Properties II

The words to the left of the colon all have a (non-trivial) property that the word to the right of the colon doesn't have. What is that property?

(a) CONQUISTADOR, CROSSBOW, DUMBSTRUCK, GUILDHALL, KILOBYTE, LANDMARK, SHRUBLIKE, STARTLING, WRITTEN: INTERNAL

(b) BEAT, CHAFE, CHIC, CLOVER, HIDEOUS, MODEL, STALK, TRIO, GARNISHED: SINGLE

(c) BRANDY, CRUISE, DECK, DOPE, EASEL, FROLIC, KIT, LANDSCAPE, SLEIGH, YACHT: CARNIVAL

(d) ASTEROID, BREWSTER, BUCKTHORN, COMPENSATE, ENTHRONE, OUTSHINE, STEWARD, XANTHOUS: LIMPIDLY

(e) ART, BUDS, COMPARE, LEASE, LOVELY, ROUGH, SHALL, SHORT, TEMPERATE, WINDS: SUMMARY

(f) CANVAS, FLASKS, FLORIST, IMAGO, MAINLAND, MAUVE, ORIGIN, TEENS, VARMINT, ZOOMING: CONNECTIONS

(g) CHAIN, ENEMY, MOAN, PLANE, REGALIA, REIGN, SERIAL: ANIMAL

(h) GRATE, DISCRETE, HOSE, PRIDE, RUDE, SEER, WEAR: PRAISE

21. Missing characters

Identify the following books from their opening lines. We've included only letters from a subset of the alphabet (a different subset in each case).

(a) -h-y'r- -u- -h-r-. --a-k --ys -n wh--- su--s up --f-r- m- -- --mm-- s-x a--s -n -h- ha-- and g-- -- m-pp-d up --f-r- - -an -a--h -h-m.

(b) I- -a- -he -e-- o- -i-e-, i- -a- -he -or-- o- -i-e-...

(c) -h--- w--- f--- -f -s: G---g-, --- W------ S----- H----s, --- --s--f, --- ----------c-.

(d) --e-l--- --l-e- ----- -is b---le f--- --e ---ne- -f --e -an-el-ie-e and -i- ----de--i- ---in-e f--- i-- nea- -------- -a-e.

(e) -ll --i- --pp-n--, --r- -r l---.

(f) - ---- - ----- ----- - -- ---- -----'- --- ---- ------.

(g) So-- o- t-- -v-- o- -y t--- --y --v- b--n -n--r-nt -n o-r --r---st-n--s.

(h) -- -h-s S--da- m-r---- -- Ma-, -h-s --r- -h- -a--r -as -- -- -h- -a-s- -f a s--sa---- -- --- --rk, a--k- m--h --- -ar-- f-r h-r ---h- --f-r-.

(i) '-ha-'- ---n i-!' -ai- ---- -e-e- -i--ey.

(j) --w-rd- -h- -nd -f -----mb-r ---7, -n -li-n -r-f----r -f ---i-l-g- -i-i-ing R---i- --m- -- --- m- in ---r-gr-d.

22. Christmas songs

(a) DKTEHSWHBSOFHYFALALALALALALALALA

(b) NLNLNLNLBNISTEKGOFIL

(c) TECSWEGTWEDE

(d) WTWLYRDYDOWNHESSYRMAKGSACS

(e) CSRGONANONFE

(f) TYNRLTPRRHJNINAYRRGS

(g) RGOTTEBSTTBMBM

(h) RTATTEFTFEBYSTASFN

(i) SGCSOFASSGINENSGALYECSOFHNAE

(j) TSSETESHADFHADAASGTG

23. Reorder I

Arrange the members of the following sets into another, logical, order.

(a) ATHLETICS, COLLOQUY, ELEVEN, KANGAROO, LACERATE, MATADOR, OVERDRIVE, PALATINE, POLICEMEN, UNDERRUN

(b) CALLING, ENTIRETY, EREMITE, ETYMOLOGICAL, IONOSPHERE, ITEM, PERCOLATOR, STATION, TEMPER, TORRENT, VISTA

(c) CLEAR, DISCHARGE, EMPTY, FAIR, FIRE, JUST, ONLY, PASSION, SUFFERING, UNDERGOING

(d) ACTION, ADORNS, COURSE, DOMAIN, ELICIT, EXCUSE, INCITE, MANIOC, NOTICE, RANDOM, ROUSED, SECURE, SOARED

(e) BEGINNING, BRUISE, DEPARTURE, FIGURES, GNOMES, LEAKS, PITY, POST, RATES, SORROWS

24. Lists II

Each of the following lists of words leads to a second list, which leads to a third list, which is in alphabetical order. What are the third lists?

(a) SHAWL, COURTESY, TULIP, CARGO, BANDIT, COACH, MOLASSES, MYTH, EDUCATION, CORACLE, MAMMOTH, POODLE, DECK

(b) STARE, BEACHES, PEAK, FOURTH, RHYME, SALTER, CARROT, BOARD, CHASED, IDLES, COARSE, CHOIRS, ASSENT, REST, PAUSE

(c) UNSTUFFY, HOWITZER, HAZELNUT, MATCHBOX, PUNGENCY, RELATIVE, SENSIBLE, SWITCHES, DEADLIER, QUESTION

(d) LATER, DOUBLER, ZEBRA, ASHORE, GUN, MARITAL, PLEAD, TRIFLE, ASLEEP, RECALL, LIGHTS, WORDS, ARTIST, TISSUE, SINEW

(e) RUGS, TRACER, TROPIC, SPICES, CAMUS, PULES, MANOR, CANTOS, RAISE, LAVE, HARDY, SCUNGY, NAMES, SCUTE

25. Where?

Where does SILENCE fit in the following list? *(Read left to right, top to bottom.)* See pages 145–8 for explanation.

PLANET	HUG	KNEES	GAME	THROAT	BUSH
FIDDLE	PEEP	PARSLEY	MURDER	DOING	DUCK
YEARS	CASTLE	HOUNDS	GUN	COMPASSES	SHANK
LIP	SIEGE	THRONE	MOUTH	NIGHT	DAY
COURT	EXALTATION	PAROLED	SHARE	NEST	SUIT
WATCH	PARLIAMENT	FLIGHT	WHISTLE	PUNCH	HORN
ART	START	TAMING	BILL	REPHRASING	BUGLOSS
LOVE	SCHOOL	CAP	HOUR	HAMMER	CROSSING

26. Identify me

My first is in a combine harvester, but not in a ploughshare.
My second is in a pigsty, but not in a cowshed.
My third does not exist.
My whole is in a farmyard, but not in a jungle.
Identify me.

27. Sums

(a) Hydroxypropyl methyl cellulose + Amaranth − Guar Gum = ?

(b) $\dfrac{\text{Ma} + \text{Mania} + \text{Moron} + \text{Sofa} - \text{Sure}}{\text{Enna} - \text{Tuns}} \times \text{Roe} = ?$

(c) $\dfrac{\text{Davison} + \text{Tennant} \pm \sqrt{\text{Eccleston}}}{\text{Pertwee}} = ?$

(d) $\dfrac{\left(\dfrac{\text{Stone}}{\text{Pound}} \times \dfrac{\text{Furlong}}{\text{Foot}}\right) + \dfrac{\text{Square Foot}}{\text{Square Inch}} - \dfrac{\text{Pound}}{\text{Old penny}}}{\dfrac{\text{Hour}}{\text{Second}}} = ?$

28. Hotels

(a) I was in a hotel recently, and in the lobby area was one of those signs which consists of a black board with lots of holes in, and a set of plastic letters which are used to form words. The sign said:

IN THE EVENT OF FIRE
OR EVEN SNOWING
USE THE EXIT

How many rooms were there in the hotel?

(b) Shortly afterwards I was in a European hotel of similar size with a lovely sea view. Amazingly there was a similar sign in the lobby which said:

SINCE THE FIRE
WHEN BUZZER HISS
EVIDENCE INN UNSAFE

In which country was I staying?

29. Longest word

Find the longest word, with no repeated letters, which, when its letters are arranged in alphabetical order has no two letters adjacent in the alphabet. E.g. DRAFT becomes ADFRT, and satisfies the criterion; WORDS becomes DORSW and does not satisfy it.

30. Order

Seven words have been enciphered using a simple substitution. Decrypt them, and then put them in the appropriate order:

FRAFR CNEGL THRFF URNIRA RFGNGR YNQL PBYHZA

31. Next pair

What is the next pair of numbers in this series?

(18, 19), (28, 29), (38, 39), (79, 80), (81, 82), (83, 84), (85, 86), (?, ?)

32.　Number sequence I

(a)　1, 2, 4, 7, 12, 20, 33, 54, 88, ?

(b)　4, 2, 5, 2, 6, 10, 3, 7, 6, 4, ?

33.　Name dropping I

The following words used to contain some famous names, before they became stars.

For example, *et cho*ol = Mark Lester (MARKet choLESTERol)

(a)	*stro sou*er	(film)
(b)	*sled re*nt	(politics)
(c)	ent* *na	(film)
(d)	*ion fri*	(business)
(e)	*h pa*e pea*	(film)
(f)	pa*ma vi*nt	(music)
(g)	cla*t	(fashion)
(h)	e*e s*y	(sport)
(i)	o*a a*	(film)
(j)	i*** ****	(music)

34.　8-letter word

Here are ten 7-letter words. But what 8-letter word do they convey?

```
R U B E O L A
  T O U R N E Y
F O X T R O T
  F A C T O R Y
N U M B E R S
  B A T H I N G
T H E A T R E
  B R I S T L E
H A M L E T S
  A T H E I S T
```

35. **Lists III**

(a) The following list of words leads to a second list, which leads to a third list, which is in alphabetical order. What is the third list?

RING, WAVE, GREASE, BURN, MEAL, BAG, PRINT, PULL, STRAP, FLINT, CAP, VARNISH, CHAIR, SHOE, FLASK

(b) The following list of words leads to a second list, which leads to a third list, which leads to a fourth list, which leads to a fifth list, which leads to a sixth list, which is in alphabetical order. What is the sixth list?

GAMBLE, SPANS, GRASPING, LADDER, STALE, STABLE, CRATES, CHASTE, RELAPSE, SENDS, PRAISE, BANGLED, SNAILS, ANODE, TROWEL

36. **Letters and numbers**

(a) What is the next (final) letter in this series:

E, O, R, X, N, T, Y, ?

(b) What property do the following numbers, and no others, share?

3, 8000, 1000000, 1000900, 2×10^{48}, 10^{62}

37. **Word sequence II**

(a) NIGHT, PAINTING, INIQUITY, THRIVE, VAPOURS, VERIFY, REVISIT, ?

(b) HIP, HEAD, LIP, BELLY, BACK, CHEST, NOSE, OCCIPUT, FOOT, NECK, ?

(c) STOP, LUMBERED, TORN, DENOTED, CONTORTS, TUBERS, NUTHATCH, BARRISTERS, DITHERS, ?

(d) HEWED, GREENWOODS, FANCY, ENGINEERED, DREAM, CONSENTS, BRUSHED, ?

(e) SUPERABUNDANCE, CALIPH, FRAGILE, ISOMETRIC, TICKLING, EXPIRY, ALIQUOT, DOCILE, ?

(f) OPENED, SECURE, SUMMIT, BEFALL, ISOBAR, LATTER, INSECT, OVERDO, SECRET, ?

38.   Film plots

Identify the following films, and give their titles and dates.

(a) Two rival racing drivers go undercover to catch a narcotics dealer, but are sucked into the drug subculture.

(b) A woman runs away to Chicago after massacring her schoolmates in the gym, and has an affair with a married man.

(c)   A blue parrot escapes from prison to kill his wife's lover.

(d) After a car salesman's brother is killed in a cattle stampede, he gets caught up in a kidnapping plot.

(e) At an exclusive school, a possessed doll is drawn into a conflict between two instructors.

(f)   Saoirse Ronan encounters a giant mutant slug monster in the Han River.

(g) A pair of assassins discover they're not really married.

(h) An ambitious coalminer is talked into becoming a boxer and builds himself a high-tech armoured suit.

(i)   Charlie Chan investigates the case of a man who gets his leg chopped off, to the theme used for the BBC's London Marathon coverage.

(j)   A wandering samurai protects a pop singer.

(k) While crossing the Caspian Sea, Jeanette MacDonald leaves Nelson Eddy and embarks on a relationship with a werewolf.

(l)   On a plane from Peru to Panama are a private eye and a nightclub singer, a fisherman married to a cannery worker, a construction worker and his unemployed pop singer girlfriend, and a group of lifeguards.

39.   Trigraphs

Explain the following sets of trigraphs.

(a) ART, BUS, CAR, COO, FOR, HAY, KEN, MAD, PIE, WAS

(b) AIA, AOU, BLJ, GHD, HGA, IJI, JEV, KCH, NDH, NGK, SSO, UJU, YKJ

(c) AIL, BUT, CAR, FIN, JET, LIT, MET, PAN, PAR, RUE, SON

(d) CHM, GHI, GSK, HMS, ISZ, KOV, KYK, LIU, LSS, UGH

(e) ANO, APH, EGA, GAN, GRA, HER, NOG, OGR, PHE, RAP, STE, TEG

(f) AYE, CHI, DIK, GRU, JUG, KIE, MUU, PUT, TAM, TEE

(g) AND, BUR, EAT, ETH, GUY, LAR, MAL, RIM, WAG, WAN

(h) ART, ASP, BEE, CAB, CAR, LEE, LET, MAR, PAR, POT

(i) ASS, BUT, HER, ION, NET, ONE, PET, RUM, TAR

(j) BEN, LAC, LED, NET, NEP, NIM, RAM, REV, SAW, SEW

(k) AAA, ANN, CIA, EAR, ETA, IDA, LAN, LOO, MOO, NIA, SAE, UAE

(l) AME, CEE, ELS, GAW, IGL, HYP, MUC, PYE, QIG, STY, UFW

And what's the missing trigraph:

(m) CDE, FHJ, GXY, HOQ, ILO, JXZ, LPT, OTY, ?

And divide the following trigraphs into two groups – one before and one after.

(n) BOK, FTR, JJJ, KOS, LQR, PCF, QDU, RFS, RHQ, TOS

40.   The list

What completes the list?

SORE

KEEL

COSH MARK

41.   Word shape

If ABBOT, DITTO, GLEE, HABIT, KELP, ORBIT, THIRD and TOOTH can be described as square words, what shape is CUFF?

42. Name dropping II

The following words used to contain some famous names, before they became stars.

For example, *et cho*ol = Mark Lester (MARKet choLESTERol)

(a) undi*uished   (music)

(b) c*nge mul*   (film)

(c) con*e *w   (sport)

(d) heat* *nt   (film)

(e) de*d c*on   (art)

(f) si*n *ock   (sport)

(g) s*le t*   (TV)

(h) s*d f*k   (music)

(i) *ry g*   (film)

(j) *a*py   (music)

43. Pairs

The elements of each of the following sets can be paired up. What are these pairs?

(a) AGREES, ASPIRE, CHASTE, ENLIST, ENTRAP, GREASE, LATTER, NEURAL, PALEST, PARENT, PRAISE, PRIEST, RATTLE, SACHET, SILENT, STAPLE, STRIPE, UNREAL

(b) ASP, BAR, BOARD, CHOPS, COME, CUP, DAM, DIES, GAIN, HALT, MILES, NATION, RAIN, REAL, REST, SIDE, TICK, TONE

(c) ALTHOUGH, BEER, BITE, BLUE, CONFINE, DISDAIN, GATEAU, LAUGHED, LYNX, QUARTZ, RAFT, RESIGN, SANE, SLEIGHT, THROUGH, TIER, WARTS, WINKS

(d) ANT, BAN, BAR, BOG, BUS, END, FOE, FUN, GAT, LEE, POT, RAG, RED, ROT, SHY, TEN, THE, TRY, URE, WIN

(e) AIL, AIR, ALE, AND, ANY, ASS, BUS, EFT, FED, FEE, FIN,
FIR, FUN, JUT, LET, LID, LUG, ODD, OIL, OLD, ONE,
ORE, PAY, PIE, PRY, RAY, RED, ROE, RUE, SET, SIT, SOT,
TIE, TOP, WAY, WOE

(f) AIL, ANY, CAR, CUR, END, FEN, KEY, OLD, PAN, ROW,
RUT, SEW, SPY, VAT

(g) ADORN, AID, AIM, AMOK, ANT, AT, BEAR, COOL,
FACIAL, FOR, GORE, HALO, HATING, IRON, JEWRY,
LAID, LARD, MANY, MOAN, OH, ON, REAL, ROAD, SANK,
SEEN, SERAPHIM, SEX, SHIELD, SNOW, SONNET, STAIR,
VIEWING, WADE, WHEN, WORK, YEN

44. Matrices

$$
\begin{pmatrix} J & A & T \\ P & T & O \\ E & S & E \end{pmatrix}
\begin{pmatrix} U & T & T \\ L & H & M \\ F & S & N \end{pmatrix}
\begin{pmatrix} S & R & E \\ E & T & T \\ I & I & T \end{pmatrix}
\begin{pmatrix} T & I & R \\ T & H & H \\ N & N & E \end{pmatrix}
\begin{pmatrix} C & X & S \\ E & E & I \\ A & G & N \end{pmatrix}
\begin{pmatrix} O & W & F \\ T & M & S \\ L & L & C \end{pmatrix}
\begin{pmatrix} ? & ? & ? \\ ? & ? & ? \\ ? & ? & ? \end{pmatrix}
$$

45. A to Z

A = SW, TS, HDS, JH, YK   N = AH, JA, PS, BH, EGM

B = JP, RDN, MP, BH, KG   O = RM, MA, SB, DL, KW

C = DA, JT, HC, TW, SG   P = TH, DW, JR, AB, JW

D = JV, BR, NB, RC   Q = PD, RW, LA, TW, S

E = HM, NT, CL, NC   R = JC, JH, MA, JB, RR

F = DW, DC, KR, JG, ML   S = DC, JB, RF, BM, JD

G = SB, GC, EH, OI, PS, AW   T = EM, EB, JLM, KM, RC

H = JLC, DP, PJS, NL   U = CE, GH, MF, RH

I = LD, KW, MC, EP, TB   V = POT, LP, JW, RG, VR

J = RS, RS, RD, LG, MH   W = YB, RB, JB

K = LA, AA, FA, LN, RC   X = ONJ, MB, GK, JS, DA

L = JC, DB, TF, CM, ST   Y = GC, PC, EI, JC, SM

M = BS, CR, DS, JPS, SBC   Z = BS, OW, WF, MJ, JV

46. Garbled text

The following text is garbled – explain how.

DCBI NE UIF MTNADSR AFVZDDM UYM COF MHOD NO
C NNCGJD QGNOF BNPSFQSMMER UN UGPFD NQ DNTP
JFUVDSQ ND UID BKQGBAFV BMF DCAI KDVUFP GM
VGHP UFZV HQ PCOENOKW QDQJBBFE CW CMNVGDQ
XGGBG AMPPFPQNMEQ VN UGD RCOD MVNADP BP
VGCV JFVVFQ

47. Cipher message

The following cipher message contains a clue to the method of encipherment:

580: 546 5603248534, 590 492

What does it say?

48. Where?

Where does HEAVEN fit into the following list? *(Read left to right, top to bottom.)* See pages 145–8 for explanation.

HOUND	ART	BLOOD	BREAD	STRAP
LANTERN	HEEL	PASTRY	MANGER	BUCKET
WOOL	WIG	GREASE	COLUMN	CLASS
OINTMENT	SOLDIER	DIMENSION	LEAVE	BELT
MATTER	SHAKE	FLASK	SUMMER	BOX
CAP	CROSS	MOON	HOLE	WOOD
BAG	PEEL	NECK	SKY	GIN
BRICK	HAND	DARK	BLADE	SENSE
FLY	ROLL	WORLD	HOLE	NIGHT
CENTURY	COLLAR	WATCH	PAGES	

49.　Groups

What group is associated with each of the following?

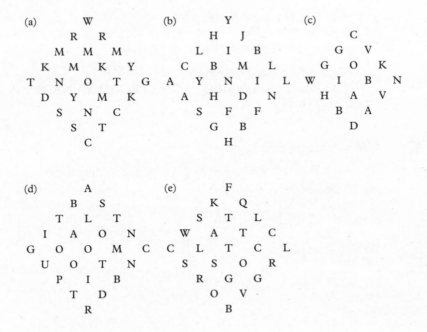

```
(a)           W                  (b)           Y                  (c)
         R     R                        H     J                              C
     M     M     M                   L     I     B                       G     V
   K     M     K     Y             C     B     M     L             G     O     K
 T     N     O     T     G       A     Y     N     I     L     W     I     B     N
   D     Y     M     K             A     H     D     N             H     A     V
     S     N     C                   S     F     F                   B     A
       S     T                         G     B                         D
         C                               H
```

```
(d)           A                  (e)           F
         B     S                        K     Q
     T     L     T                   S     T     L
   I     A     O     N             W     A     T     C
 G     O     O     M     C       C     L     T     C     L
   U     O     T     N             S     S     O     R
     P     I     B                   R     G     G
       T     D                         O     V
         R                               B
```

50.　When was that?

The author of that favourite novel of yours? Well he shares his surname with another who came from Sheffield and who wrote a novel set in a fictional seaside town that shares its name with a real one in Devon. That town has a castle that isn't a castle, and the architect of that worked with another architect in Plymouth and this second architect died on the same day that an author first gave a lecture on manners. He gave part seven of the series on the same day as someone else became famous for surviving. The husband of the artist who painted this event wrote the biography of a man who was killed in a battle that ended a war. The peace treaty was signed on the same day as two Nobel Laureates were born. The one of these who wasn't a writer was born in a town where a famous conference took place and the man who presided over this had been crowned king only recently in a cathedral that was built by and is the resting place of a man who was authorised by a pope to take anything he wanted from an Italian city where a Roman emperor was born and where his daughter married the king of a tribe who was previously married to the daughter of a king killed in battle. In which year was that battle?

51. Sums

(a) Vinegar + Carbolic Acid + Carbon Dioxide – Aspirin = ?

(b) $\dfrac{\text{Nebuchadnezzar} \times \text{Methuselah}}{\text{Melchior} + \text{Balthazar}} = ?$

(c) (Baseball × Cricket × Netball) – (Basketball × Rugby Union × (Rugby League – Polo)) = ?

52. Number sequence II

(a) 7, 14, 19, 29, 40, 44, 52, 59, 73, 83, 94, ?

(b) 18, 26, 38, 62, 74, 102, ?

(c) 18, 25, 28, 34, 35, 37, 41, 44, ... (44th element)?

53. Chains

(a) SOYLENT   ____   CARD

(b) DIE   ___ ___   SQUARE

(c) UNIVERSAL   ___ ___ ___   MAGNOLIAS

(d) CABIN   ___ ___ ___ ___   MAN

(e) NEAR   ___ ___ ___ ___ ___   INTERRUPTED

54. CEB VRD CPC VPB?

CNT VOD VIE / CTC VSD / CEK VUD CQP / CSK VIB VHB CTC / CNP VID / CNA CZN VCB VSA CEH / CDP CEF VCB / CEB VRD CPC VPB / CUP VOD / VRA CGK CRO CER VTB CTN / CEP CLF VEC CEO / CRM VHE VTB CHO CCP VIC VHD CWK ?

55. Missing letter

What is the missing letter?

?, ?, ?, ?, n, ?, ?, ?, ?, ?, ?, ?, u, u, i, n, i, o, ?, u, n, i, i, i, ?

56. What doesn't come next?

Here's an antidote to 'what's the next in this series' questions.

(a) What ISN'T the next state of the USA in this series (of those remaining)?

INDIANA, ALABAMA, COLORADO, SOUTH DAKOTA, NEW YORK, CALIFORNIA, GEORGIA, MASSACHUSETTS, WEST VIRGINIA, NEW JERSEY, ALASKA, OKLAHOMA, MONTANA, VIRGINIA, VERMONT, MISSISSIPPI, —, ARKANSAS, KANSAS, TENNESSEE, UTAH, NEVADA, DELAWARE, TEXAS, MARYLAND, ARIZONA, LOUISIANA, NEBRASKA, NORTH CAROLINA, IDAHO, KENTUCKY, FLORIDA, OREGON, MICHIGAN, ?

(b) What ISN'T the next African country in this series (of those remaining)?

LIBYA, CHAD, SOUTH AFRICA, MALI, GABON, RWANDA, ALGERIA, MALAWI, GHANA, BURKINA FASO, MAURITANIA, NIGER, GUINEA, ZAMBIA, TOGO, LESOTHO, TANZANIA, BOTSWANA, SUDAN, UGANDA, NIGERIA, SWAZILAND, ERITREA, ETHIOPIA, ZAIRE, BURUNDI, KENYA, ZIMBABWE, ANGOLA, SOUTH SUDAN, ?

57. Jeopardy

(a) Judge Doom

(b) One joined the army and one married Thelma

(c) It makes none

(d) Because she would have known that the old man was an impostor

(e) It's behind a double-parked van

(f) She lives in a mansion with her sister Blanche

(g) Put him in the long boat and make him bail her

(h) Joff-tchoff-tchoffo-tchoffo-tchoff!

(i) 36°35'N 98°28'W   32°43'N 117°10'W

58.   A poem

Whilst driving recently my car felt weird.

The steering failed and off the road I _____.

A gentle stop was not on the agenda.

I bruised my leg, and gosh did that feel _____.

I _____ right by the vehicle till the AA came along.

They spotted that the brake alignment had been set up wrong.

The mechanic in the garage said he would have to _____

and then replace the brakepipes, back into their groove.

He tested them and it turned out that the test results were poor.

So I asked for a _____ just to be quite sure.

You'll see I've left 5 words out of my little doggerel rhyme.

They all have 6 letters – which should save you lots of time.

These words all fit together in a sort of square array.

Use this to solve the question: This all happened on what day?

59.   In common

What do the following have in common?

ADIER AIN ERAL EREAN IRAL OMMANE OMMOO ONEL OR
ORAL RIVA

60.   Reorder II

Arrange in a different logical order.

CROSS, DOT, DROP, MIND, ROLL

61.   ... and in English?

In French only (2, 3), (9, 10) and (10, 11) have this property. In
German only (4, 5), (5, 6), (7, 8) and (8, 9) have this property. What is
the situation for English?

62. Where?

Where does COSY fit in the following list? *(Read left to right, top to bottom.)* See pages 145–8 for explanation.

SCARE	PRINCIPLE	YOURS	CONSTITUTION	SUN
GARDEN	FIFTH	HOPE	GLOBE	SNAP
BANG	BLUSHES	CHINA	PRESS	BURST
BEAN	DISPATCH	PAR	HORN	PLAN
STONE	TEACHER	CHOICE	LAMP	CHIN
SPINNING	GO	BLACK	QUESTION	THANKS
PROJECTION	HERALD	SLIME	TOOTH	REVENGE
BOX	TRIANGLE	BLUE	LOPE	CHECK
ONE	ROUND	DOG	CUBE	RUNNER
BEE	PLOUGH	LAZY	BIRDS	TODAY
POST	FOUNTAIN	SWEET	SOCK	DRAYS

63. Sums

(a) (Monkey + Dragon + Rabbit) – (Dog + Rat) = ?

(b) (OF × EA – FR × SL) / (CR + AL + SH + OZ) = ?

(c) (Hobbits in the Fellowship + Elven rings + Dwarvish rings + Towers) × Black Riders = ?

(d) ((Anna + Puss – Liss) / Nerk) – Pier = ?

64. Where? – with a difference

Here is a slightly different Where? question. Normally we tell you which word you need to position in the list. Not this time. Not only do you have to work out where the missing word belongs in the list – you also have to work out what the word might be! *(Read left to right, top to bottom.)* See pages 145–8 for explanation.

JANE	FOLLOWER	CITY	PET	DATING
SINGER	WHEEL	DICTIONARY	GUEVARA	MATE
GRATER	RUN	SUEY	CAROL	HANGER
SPANIEL	WORKS	PUNISHMENT	PIE	POTATO
BOY	SAUCE	CARD	RECORD	CIRCLE
OIL	BALL	ROOT	FRAME	BOARD
SHOP	AFFAIRS	FISH	SARK	

65. Chessboard

(a) Whilst tidying the Kwiz vault we found a chessboard on which each square had been marked with a number. The top right corner of the board had the following markings:

4 5 4 5
3 4 5 4
4 3 4 5
3 4 3 4

How were the squares in the bottom left of the board marked?

(b) Another chessboard had the top right corner marked:

5 6 6 6
5 5 6 7
4 6 6 7
6 6 6 6

How were the squares in the bottom left of the board marked?

66. Identify and divide into pairs

(a) Band from Wigan

(b) Word that is the theme of this question

(c) Surname of former Foreign Secretary

(d) 1961 film that marked the final appearance of two Hollywood legends

(e) Novel by Stephen King

(f) Band who debuted in May 1979 supporting Scritti Politti

(g) Best picture Oscar winner of the '70s

(h) European city

(i) ASBO superheroes

(j) REM single

(k) Singer, born in 1951

(l) Band from Wigan

67. Complete the set

5, 8, 9, 12, ?, 20, 23, 25

68. 63 words I

The 63 words below can be divided into 21 sets of 3, with the 3 words in each set being linked in some way. Of these 21 sets, 20 of them can be paired into 10 pairs by means of a connection between the sets – perhaps a common theme, or the sets have been formed in the same way, or the links themselves are connected in some way. Having removed these 60 words, which 3 words are left, and what is the link between them?

A	ANNE	ARTHUR	ASCERTAIN	AUCTION	BANK	BARN
BELL	BLENHEIM	BOW	BOXING	BRITAIN	BURNS	CARTESIAN
CHURCHILL	CLEANER	COURAGE	DANCING	DARWIN	DOG	DOWNING
ELVIS	FIELD	FLY	GABLE	GARRIDEB	GET	GROUNDHOG
HAT	HIS	INTEGRAL	IRON	KONG	LAST	LINCOLN
MAIN	MAN	MAT	MAY	MUSTARD	ORIEL	OUT
OVERS	PEACOCK	PLUM	POT	RELATING	ROME	SECTARIAN
SHOPPING	SILENT	SILL	SLUYS	STEP	STEPS	STUDENT
TO	TRIANGLE	TWELFTH	UNIVERSITY	VERSAILLES	VICTORIA	VIENNA

69. Identical property

Divide the following lists of words into pairs, with each pair having an identical property.

(a) BALLS, BATH, BLACK, BOOT, BUTTON, CHAIR, COLD, GREASE, IRON, NUT, PADS, RIGHT, RING, ROSE, SECOND, SPRING, TEA, TENNIS, TIPS, YELLOW

(b) BLOOD, BOTTLE, BREAST, EARL, ELEPHANTS, GERMAN, HOUND, HOUSE, JET, LINCOLN, LINING, MARKET, NAVY, PEEL, RACING, SEEING, SNOW, WASH

70. Two things in common

The answers to the following clues have two things in common.

- Bicarbonate
- Alien spacecraft
- Toy figure
- Relations frequently the subject of jokes
- Authorizing official (especially US)
- Items of ladies' underwear
- Credit card
- Delays action
- Reinvested
- ?
- Silhouette
- Ability to sustain effort

(a) What are the two things the answers have in common?

(b) Add another possible answer.

71. Lists IV

Each word in the following lists leads to another word, in a manner which you are to determine, and this second word leads to a third word, in a different manner. These third words are in alphabetical order. What are they?

(a) ALLEY, DESERT, POLAR, CRY, HOT, LOAN, MARCH, SEA, LEMON, SCREECH, SHIRE, ROLLER

(b) THE HAGUE, VALETTA, ANKARA, MADRID, HAVANA, PARIS, LISBON, COPENHAGEN, BERN, MOSCOW, NEW DELHI, BEIJING

(c) CENT, HEAVEN, FOOL, HAMLET, ATTITUDE, MECHANICAL, OUTRAGEOUS, DIOCESE, VOLITION, BRASSICA, RIPE

(d) RICH, HAIRY, WARM, CROOKED, ALIVE, DARK, UNWELL, SANE, LOW, UGLY, FAT

(e) MOCK, BED, WAG, TELL, WRAITH, FROM, TOP, METRE, NAME, RUNG, BOOK

Puzzle Hunt

These next pages contain something a little different. A Puzzle Hunt is a collection of puzzles, typically without instructions. Figuring out how the puzzle works is in itself part of the puzzle!

Each year GCHQ staff gather together in teams of up to four, to try and solve all the puzzles within two and a half hours.

The answers to the first 8 puzzles are all words or short phrases.

The final puzzle is called the Meta Puzzle. Meta Puzzles typically require you to have solved all or most of the previous questions before they can be tackled.

Good luck!

Hint: The answer to the Meta Puzzle is an eight letter word.

Also in this plate section you will find the competition announcement. The Sudoku, and some other puzzles from this book, can be found at www.gchqpuzzlebook.co.uk, in case you wish to print off a copy to ease solving.

Used exclusively on the dark spots on non-sweet potatoes

Survive while many perish

The Heavens are going to get more and more crowded

Pottery
firing
permit

★★★★☆

The leaf
gatherer
is always
complaining

★☆☆☆☆

☆☆☆☆☆

One of
Thomas
and his
calf is
immortal

SHIKAKU

Divide the grid along the dashed lines into a set of rectangles so that every cell is part of exactly one rectangle. Each rectangle will contain just one number, which must indicate the total area (in cells) of that rectangle.

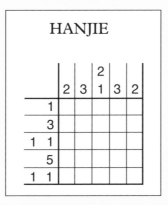

HANJIE

Paint a picture by shading some of the cells in the grid black. The numbers on the outside of the grid indicate the lengths of the consecutive shaded segments in each row (from left to right) and columns (from top to bottom). There must be at least one unshaded cell between each of the segments.

TAKEGAKI

Draw a single loop, made out of vertical and horizontal line segments, between the dots. The loop should never cross itself. The number inside a cell indicates how many of the four possible line segments surrounding that cell are occupied by the loop.

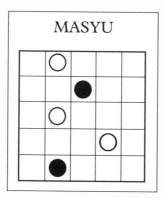

MASYU

Draw a single closed loop, travelling horizontally and vertically between adjacent squares, that passes through all the white and black circles. When the loop passes through a white circle, it must proceed straight though the circle but turn immediately in at least one of the two adjacent squares. When the loop passes through a black circle, it must take a 90-degree turn, but it cannot turn in the first square immediately before or after the black circle.

4. I Spy (8)

A deity (6)

An absence of light (5)

A shambles (5)

A container (2)

A recursion (7)

A shelter (4)

A scholar (7)

A county (4)

A drink (5)

A musical instrument (4)

A microorganism (7)

A capital city (4)

A social prohibition (4)

A material (4)

A Caribbean cult (5)

A marital status (6)

An American tribe (4)

A mode of transport (6)

A type of hat (2)

Note: There is no letter Q.

B3A5CD50B54E28240F0EBB72BD1F22528BBB208EFEE272A0D4704E966D809B812
493FF6C3C9BF09C21DDD9FBFC08D71803591AA7924315DEA1ADE3824B3535B93C
F28425D66C4C3BE9B74EE768F9B317D0329BF9ACB636BA579E3BE00B2590A5739
20322AC05BE45FED6D5075E4DB0D7018D7F5D3F66D85DF7DAK843FA83EE97ECD6
9EFC6FD516228DA0B6E9E33EEB872F34E0A38014C8700L1FFC66AD96D7ADEA414
A314A3B07DA61E31BA0B636ED96B6CF9A2545B8B16DABBC09A14BF3E1BJ5FE94C
CBD4BB4955B1CA0C2C13A4AB0BD35BF0081FF06B8EEE9AMF4BBE972F64DCC3B3E
16BDF36C237E5813A0C43AD8EEC5C9B40A6CC98F0EBDD3EA6E68125466FIA5C89
4900ED1909CD8D4613E443C79D624354AF6064B55CEC55D20E41D6HF471AD99BC
9C0BA696F144AE152C1B47C44A3C33CA174A3D0D556F99AD8FE951A39DC75BEA8
ABECA2CCDC583CE5A6D0C4OB60F1F41FEC20CC176D80B5CE5BCD6AD46A9B21E50
EA31B04659D9485666D397FCE535A17ED937EFAD408113AEF8D6E00690D60ED23
7129F345361F139C368CB3FABF1C32FE95D890CCN0EDC2C3FEB819B27FA56F7F4
8EC2F42FDF7D4ACE4D8969E4D4C1B8222279AAE2A0ECY653C3E66185B1312200D
11440212B8BBD8C5D03D332C040E546CFAC5E1AD9041C92G0DDE5C9BDABC0972B
3D5FE1FE17B29506257C1EE588B987C88091FP0DAB69E12ECD1B0906926EDCC83
55A116309B1090B13A084217F1AB681706DBAEBA4673E9CB35B2DB268409C4133
0CD1D49ABA5107278351803BDD1E1528D7EDACDE0D1W26603EC42D859DED9DAA1
64F84AF2FCB3D03A069964ADF6F8E2FF1CBC1B6AEA69F385E3BF3DC7422245645
B90C2FE4B10F64BF30E8FC9683878312A037C1B561747131B356FDEA0510X4A49
5775590521F754E32741CB40E01481E9120FA2836619AFFF67E5E380A5DDB3CFF
E68CD29D0967D28E54FE32823B7C255D0D682648CB2F1CCBC76031FE15B1415DC
9B678FB74C155A5B4042D603EAD10B7C84BDFC8A3AF56188C0E6907D9BB6936A1
5BB8492105CD1BBDB902C11EFC40649033DA1FDBA9444E0E510FBC00573BCC213
4BB29BE7FEBB7F58A7FB329B2E40F69113FABE1D2028DB2563D34F867C5476B2A
7C2074F8752FDFB695B0E58BDR48104B4B3828FD10BA877BF5E486C4C0B29AB31
30EC1FC84711C50D5A2ABF473186748AV112EEE08BFF906B81A7BB844C97C85C3
9F683C85F150109A86C7649C2AA4DE9C0BC125269D63C44C95275CC97174FA3B6
7201CC09C4E1781B3B58DF2FC2B1C8EFDD050693279239EF3A27C374FA26F2096
0A09BE17E915CE3C452A0B6E096F43EB0E02D2C2652E35B9F0F2BED6FE2C22753
1F4CD79B5301153DFC6CB8E18227920D09836BD2AF584DD888A6A03851A972AD9
4E639DCDA3C0DC196C504AE515B134DAAADC92DCBB525A09AC502E65A6BCFE90E
3F923CCE451719FAF9CEBECD6E8CA537D450DAB8FC129002F61ABD47E15B557AF
40C8FF8C4C77EDB84481B2B6F289C57F22B930F81C3AE4B7653F9CA91A3EB0E3E
2FEFD25830A48FBB1B3584B839A65A9EFFA547D49DD8C3D401CE0D0DD9C0DD4BA
7C8263153A19873D4B0F6F69531F68C0BD4E7E7CE87D49821F7A16BEDCDBBD405
708BF3F8A6F6E7DF7BBEE61947DB2C69D3B6EB0B6C98D3E4261EEDACA4C48B09B
7DF7C60F78S626EB1AE9CAB0B90CC5559735E0848A91116955D3FC07FFABD619F
9C3183C9890E88CE83593C3ED071AAF4906EB2BC40EFBB4436219EDB3620916DE
BF9DCBFDA6A38A2EDU53E89B7F076E089E2B1D1BE5AF9FFEC24412019A93DCAB1
3598E6D73AEZ628A694F637BE114C214B45FC8B7D94A2465F024A3EDAE4BDE8E2
430B7F5AB9F85FAD2B9EF8B688256F22691275E10C1AEE547CC02684534134189
76115354ABDE093AF0D98EC9CD45925CB1D6C243D3F2F16BEEE484678BE795A9D
AA037350FA11DB6040A29C518F4F9850E56C472867CA103F74346936765699A93
91F09TA00D5D8281278AD7402992CC719CE20A01A296146B9C38ACEDDA24684F1
CCBD3B42B8B66D5F827B06A9BB20900D764416B8F3A12B5A132EB6CB61BEF6D03

6. Countdown (4, 7, 2)

TSOIHTALIISOWORTHREEUTWOYONEBTRIIDUOONUBKOL
MEAKAKSITYKSIUSAMITORIRERTIIDREIRZWEIDEINSVUK
UEBIYUND'AYAEWEJDCHA'NWA'OSETUDULSHANAASAMNS
ONGMNUENGNTIGAIDUABSATUISANCERHYIITREEEJEESS
NANEETORURRUADTAHIOPXEY'MUNEY'AWETRZYKDWAS
JEDENSKINSAAISKAYDHUQDTEHEBAHXHIKUTRESLDOSA
UNOBNTATHUAMBINIPNYEKHA'INOKWILINWTOPINTE

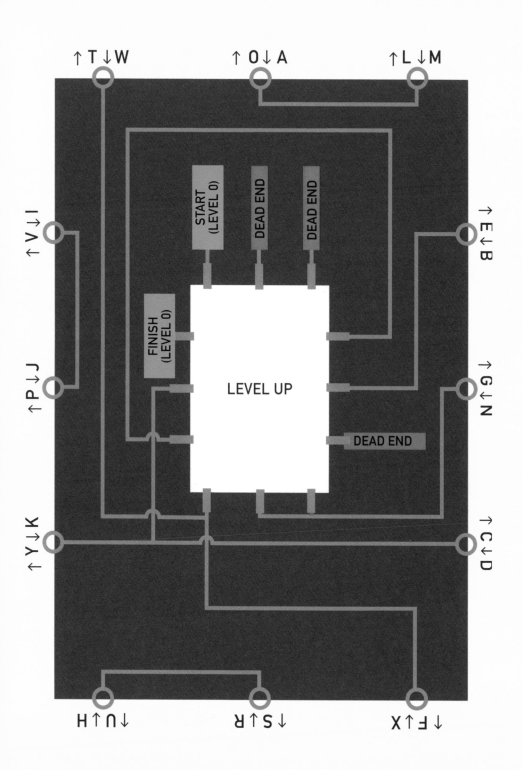

Meta

Which answers contain the letter O twice?

Which answers have no letters in common with GEEK?

Which answers contain the bigram AT?

Which answers begin with the same letter?

Which answers contain an R in the 2nd position?

Which answers contain an A and an N exactly once each?

Which answers contain at least one letter not found
in any other answer?

Which answers contain more than one word?

0.666...
○

0.222...
○

0.75
○

0.5 ○

○ **0.5**

○
0.538461...

○
0.375

○
1

Competition

Following the popularity of the GCHQ Christmas card Competition last year, here is another competition to test your mettle!

Like the previous competition, this one has various parts to it, and these will require both puzzle-solving and codebreaking skills. The first part of the competition is below, and is in the form of a Sudoku. Solving this will lead you to the next part.

As you proceed, the instructions for entering the competition will become clear. The competition closing date is 28 February 2017.

Good luck!

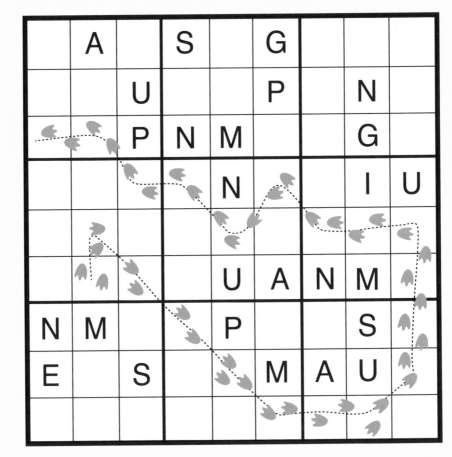

Identify the
Following People

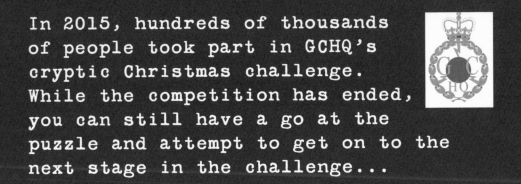

Instructions:

In this type of grid-shading puzzle, each square is either black or white. Some of the black squares have already been filled in for you. Each row or column is labelled with a string of numbers. The numbers indicate the length of all consecutive runs of black squares, and are displayed in the order that the runs appear in that line. For example, a label 2 1 6 indicates sets of two, one and six black squares, each of which will have at least one white square separating them. Complete the grid carefully with a black pen and check your answer is complete and correct before proceeding.

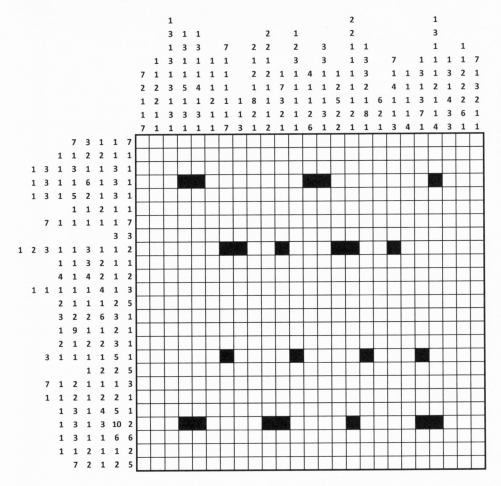

72.   Word sequence III

(a) QUARTER, DUTY, LINER, POSTER, OFF, ASIDE, PINS, STEP, ?

(b) ALERT, AGENT, BELLY, LAND, BIRD, MOOD, ?

(c) ELOPE, BUG, ENTER, REIN, LATE, OUT, SUCKER, ?

73.   Name dropping III

The following words used to contain some famous names, before they became stars.

For example: *et cho*ol = Mark Lester (MARKet choLESTERol)

(a) se*e ma* nar*psy   (music)

(b) tar*in par*y   (music)

(c) ho*ss   (sport)

(d) diplo* stin*   (music)

(e) r*mp cha*e   (politics)

(f) *i go*k   (actor)

(g) s* va*   (actor)

(h) na* p*   (astronaut)

(i) *b t*   (actor)

(j) r**   (film director)

74.   Reorder III

Arrange in a different logical order.

587 YARDS (approx), AGAINST, CARBON, COMPACT DISC, RALPH FIENNES, GREEK LETTER, KISS, LIFE STORY, ME, SRI LANKA, VIOLET

75. Morse Christmas music

Here are the titles of some Christmas pieces, given in Morse. To keep things simple, we have shifted all the spaces to the right-hand end of each line. Identify the pieces.

(a) ......-.-------------.---.-...........-.--.....--.-.--.-
..-...-..-...

(b) ..-.-.....-....-....--.---..-.-...--..-.-.-.

(c) --..-.-.-.---.-......-.......---.-...-...-.-..-....-.--.

(d) ......-...-.--...--.....-

(e) .--..-.-..-..-----.-...-..-...--.-..

(f) .-....--.-...-..-...----..-.-...---.--

(g) .--.--...-......--.......-...-.........-.-.-..----.......--
.....

(h) ..--------------------...-.......

(i) .--.-.....-...-.-..-.--....---..-.---.-.....-.-.-.-..

(j) .-...-..-...-.---.--

76. Properties III

The words to the left of the colon all have a (non-trivial) property that the word to the right of the colon doesn't have. What is that property?

(a) BOND, FACE, FRET, GIST, LACK, LYNX, MONK, ONCE, PUNT, SOUP: QUIZ

(b) ANKARA, BEIJING, BERNE, BRASILIA, CANBERRA, NEW DELHI, OTTAWA, RABAT, WASHINGTON, WELLINGTON: LONDON

(c) ACROSS, ANTEROOM, APPALLING, ARROGANT, CARRIES, CHAPLAIN, CLOTHED, FLOGGING, INCOHERENT, MALICE, PROVABLE, RAIDING, UNDULATED: AXILLARY

(d) AMBITION, DIPLOMAT, ENDORSABLE, FURTHERMORE, MANICURE, NOBLEWOMEN, ORIFICE, PIMENTO, RUMINANT, SOARING, TOEHOLD, WEATHER: SKELETON

(e) ABSTRACT, BASS, CONSOLE, INCENSE, PRESENT, REAL, REFUSE, SECOND, SEWER, SLOUGH, TEAR: QUESTION

(f) AARON, ALEXANDER, ELISABETH, EMMANUEL, FLORENCE, GERALDINE, GREGORY, LEONARD, THOMAS, TIMOTHY: ROSALIND

(g) APPEAR, BRANDISH, CLOWN, COLLAR, CORRECT, FALLOW, GROVEL, LEVER, LURE, RUSTLE: DIRECTIONAL

77. Odd archive

An archive of film, music and literature has been catalogued by someone who wants to keep entries as short as possible. So FOURXKYRNFHE was written by Richard Curtis, RATBGR was by U2, and THEPRISONMSGXZPZUVFF was written by J. K. Rowling. What are the usual titles of the works recorded below?

(a) THGNOQLDOVXTMC

(b) W

(c) PREYXXALC

78. Name

M lives in Q, and is very fond of G. He is married to J, and they have two children, C and V. What do the children call M?

79. Odd one out?

ABBREVIATE, CONTEMPORANEOUS, DEGENERATIVE, FLOODWATER, FOXY, GRAPHICAL, INTERMEDIATION, JIGSAW, LIQUIDS, UNCOUTHLY, ZEKS

(a) Which word in the above list is the odd one out?

(b) Why, actually, isn't it the odd one out?

(c) If you were to make a similar list of words, but with a different number of words in the list, how many words would it contain? (*No need to make such a list, unless you want to!*)

80. Miscellaneous

(63°09'N 117°16'W 37°55'N 120°25'W 02°07'N 18°06'E
56°12'N 3°0'W) – (37°07'N 82°36'W 14°30'S 69°18'W
51°39'N 02°27'W) ?

81. Peculiar event

What is the last word in the following?

SOMETHING PECULIAR HAPPENED WHEN I SAT DOWN
RI YSW RHW RTOWQEURWE. RHW FAYLR GIR QIESW
YNRUK RGW RWXR CIYKS NIR BW EWCIFNUAWS. UR
AWWNWS RIVW XINOKWRW FIVVKWSTFIIJ.

82. A to Z

A = T, U, T, C, P, I   N = N, C, C, B
B = I, M   O = Y, SA, ???
C = P, B, A   P = S
D = E, E, S   Q = SA
E = C, P   R = U, T, B, DROTC
F = B, L, G, S, I, M, S, A, S   S = A, H, C, I
G = A, M, B, T   T = L, A
H = A, S, U, R, S, C, S   U = B, A
I = F, S, A, S   V = G, B, C
J = {}   W = -
K = K, C, T, U   X = -
L = A, S   Y = SA, O
M = U, R   Z = Z, M, SA, B

83. Odd one out

Which of the following is the odd one out?

(a) AIR, ART, CHI, DOG, HIS, ITS, MRS, RAM, SIT, TAR

(b) BIY, FEG, FEL, FER, HOM, JEG, MED, PON, SEP, SUT

84. Round Britain Quiz I

This style of question commemorates Round Britain Quiz *on BBC Radio 4: the format is one, long, cryptic question which has six parts to it, indicated by the letters a–f. To gain full marks you should identify all six parts. This will be sufficient to answer the question.*

The following individuals are arranged in approximate order of size, starting with the biggest: a fictitious Australian who moved to New York (a), a winner of 14 majors (b), John Wayne, in Oscar-winning form (c), Yusuf Islam (d), an altruistic bandit (e) and a former speaker of the House of Representatives (f). How is this?

85. Next in sequence

(a) T, P, O, F, O, F, N, T, S, F, ?

(b) Find as many as possible of the next five terms:
      6, 4, 40, 12, 70, 56, 36, 100, 33

86. Riddle

A riddle.

My 1st is 2nd.

My 2nd is not big.

My 3rd belongs to us both, in a sense.

My 4th to itself is humdrum.

My 5th sounds unconvincing.

My 6th is in Belgium on the coast.

My 7th is yes, right.

My 8th has rotted away, you might think.

My 9th is now 21st.

What is my 10th?

87. Quotations

Below are six quotations, each enciphered using a different substitution alphabet.

(a) DOKEY BR EOKEY, GUT IBPULQ BR PUBDFNQ.

(b) IHEW BP M IHHK VMED PJHBETK.

(c) D COH NRPDPQ RURNYQLDHI RXCRKQ QRGKQOOQDJH.

(d) CSAQ, JBFGAJ CSAQ BGJ QTBTSQTSNQ.

(e) MLR HRF YK CYULMYRJ MLEF MLR KSGJA.

(f) NJDTB NIRPCY TGO ETMMY T HLS NPLEB.

(i) Who originated each quotation?

(ii) Encipher the first three words of the quotation which ends 'ALL THE PEOPLE, SOME OF THE TIME'.

88. What's out?

If RIGHTS becomes ZONAL

APPEAR becomes PETER

YOUTH becomes MINA

TWIST becomes PEAL

DEEP becomes HIT

JUGS becomes DAY

What does OUT become?

89. Film sums

(a) Airport × Chowringhee Lane / Men and a Girl = ?

(b) (The Private Life of Henry × Richard) + Pola – (Phase × Henry) = ?

(c) I Died (Grams False Step) = ?

(d) (Les from the Crypt × Nan the Barbarian) – (Slight × All Change) = ?

(e) (Hoes of a Summer × Fence of the Realm) + Male Vampire – Ending Your Life – Ward Scissorhands = ?

90. Where?

Where does MUSTARD fit in the following list? *(Read left to right, top to bottom.)* See pages 145–8 for explanation.

DENTIST	BARBER	BEER	CAP	MATCHES
RING	BUSTING	WORMS	GRANT	WINE
MUSICIAN	WIRE	PARSLEY	SHOEMAKER	ASTRONOMER
DAY	SOUP	CIDER	FIREMAN	BOOTS
WILLOW	BAG	CARRIER	STICK	TAILOR
BOX	HEELS	CAKE	JAM	CANDY
CARPENTER	PRISONER	BALL	WEST	CIGARETTES
BACK	PLEASANT	IRON	AGONY	CLOVER
NECK	DRAW	CROP	FIELD	DIP
FENCE	WRIST	SHALLOW	ELBOW	TOOTHPASTE
TROT	HEAT	PRICE	CULTURE	HAY

91. Letter sequence II

(a) E, D, C, F, T, ?, Y, H, N, J, I, ?

(b) T, B, O, N, T, O, E, H, T, S, H, ?

92. Number sequence III

(a) 1, 4, 8, 13, 21, 30, 36, 45, 54, 63, 73, 85, 95, ?

(b) 4, 5, 8, 8, 9, 9, 12, 13, 13, 13, 17, 18, 21, 22, 22, 23, 26, 26, 27, ?

(c) -2, -1, -2, 0, 1, 3, 2, 3, 5, 7, 5, 6, 5, 6, 8, 9, 8, 10, 11, ?

93. Identify the following

(a) LG=SJAG

(b) MH=MGZM

(c) AR=HAFO'B

(d) HL=HCAKzS

(e) JF=JdBdH

(f) JR=JMyH-J

(g) JW=JWPLBH

(h) KB=CIEMIdF

(i) RV=RARPFGdVd'A

(j) EQ=FD&MBL=DN&EBL

94. Where?

Where does SWINEHERD fit in the following list? *(Read left to right, top to bottom.)* See pages 145–8 for explanation.

UNINTENDED	OCTAVE	SIXTEEN	OBSTRUCT	DEEP
NARCISSUS	LAGOON	JACKIE	MIST	LADY
GARDENER	JUGGLER	COATING	DANCERS	PERFORM
FLORIDA	LEVEL	LORD	FLOWERS	FRUIT
ARMFUL	BERETS	FOUR	BEHOLD	SILENCE
TWO	LUNCHEON	MYSELF	HARVEST	EIGHT
LAST	RIVER	HORMONE	NORMAL	TWENTY
CLOCKWORK	MEADOWS	COLOR	CONCERNING	SIX
UMBRELLAS	HEAR	NAME	TWELVE	ACUTE
HIKERS	THEREFORE	LATE	HEAT	TITANIUM
THREEFOLD	DAY	WIND	CHRISTMAS	WAR

95. Missing series

Which series are missing from the following sequences?

(a) OR, ALLY, STRESS, BLED, LACE, TENT, THE, ZEN

(b) 1, 4, 9, 6, 25, 6, 49, 4, 81, 0

(c) BATE, ADDRESS, CENT, PROD, NOBLE, PATH, UNIFORMED, NOSE, ACTION, BEFIT, DOTING

(d) DIAL, NOSE, STERN, CHINE, ICE, USAGE, GENTLY, STING

(e) DOOR, HOST, CHAT, RENT, HATE, ANER, SEED, RUAY

(f) PANT, SEER, MAIM, TIED, SETS, CARE, BOBS

(g) ONTO, EAT, OUT, ENUFF, ROOF, EYES, XIS, NURSE, TRAIN, EAST

There are two possible series missing from this sequence. Find both.

(h) HER, LOVE, RAT, TAN, AM, PEA, ASTER, LOT, AT, STARLING, VALE, NET, EN

96. **Who are they?**

    (a) J. S. Brook   (e) Henry Churches

    (b) Bernard Long   (f) Beautiful King

    (c) Francis Jackdaw (g) Jacob Newhouse

    (d) Joseph Green   (h) Boris Parsnip

97. **Properties IV**

The words to the left of the colon all have a (non-trivial) property that the word to the right of the colon doesn't have. What is that property?

    (a) ANIMAL, DELIVER, DIAPER, LAGER, REBUT, SPACER, STRAW, SUNG, WARDER: LEPERS

    (b) ALARM, ANGLE, DROVE, IDEAS, LEASE, LOYAL, PASTA, RAISE, SONAR, TOUGH, VERSE: AIMED

    (c) CLONE, COFFIN, COLLAR, CLEMENT, CURIOUS, CREASE, CARRIAGE, CARTRIDGE, CRANIUM, CRINKLE: CONTEST

    (d) FLOWING, GOLDEN, HEAVY, HIDDEN, INACTIVE, MOON, NEW, STENCH, STONE, STRANGER, VIOLET: ELEMENTARY

    (e) ARISE, COIN, DERIVE, DIED, IRATE, LIT, RALE, SAID, SHOED, STILE, TAMER, TILE: VOLUME

98. **In the archive**

Whilst delving in the Kwiz archive we found a piece of squared paper on which someone had written out a set of letters several times in slightly different ways. A number of 3-letter words were highlighted: three animals (EMU, FOX and GNU) and a girl's name. What name was it?

99. 63 words II

The 63 words below can be divided into 21 sets of 3, with the 3 words in each set being linked in some way. Of these 21 sets, 20 of them can be paired into 10 pairs by means of a connection between the sets – perhaps a common theme, or the sets have been formed in the same way, or the links themselves are connected in some way. Having removed these 60 words, which 3 words are left, and what is the link between them?

ALAN	AND	ANDREW	APE	ARMS	ASSET	BACK
BANE	BE	BOX	BUSH	CARTER	CEASE	CHAIN
CLOTHES	CONNECTION	COPENHAGEN	CRAIC	DAY	DRY	EDEN
ELEANOR	ETHER	FEAR	GUINEA	HOOVER	IF	IGNORANT
INDICT	JAY	KEEL	LANDS	LUMP	MADONNA	MAJOR
MICHAEL	MILE	MUSIC	NAPOLEON	NEW	PAGER	RANDY
SANK	SEATS	SEX	SHALL	SITE	SLACK	TESSA
THATCHER	THEN	TIGHT	TITO	WARDEN	WARNED	WAYNE
WE	WHEAT	WHEN	WRITER	YAK	YELLOW	ZEAL

100. Idiomatic journeys

(a)	43° 42'N	86° 22'W	to	34° 23'N	102° 07'W
(b)	52° 36'N	0° 11'W	to	52° 19'N	1° 09'E
(c)	51° 15'N	1° 07'E	to	57° 22'N	4° 12'W
(d)	0° 42'N	116° 38'E	to	52° 28'N	8° 09'W
(e)	52° 23'N	11° 28'E	to	4° 33'S	129° 41'E
(f)	35° 17'N	91° 57'W	to	all all	
(g)	36° 56'N	121° 47'W	to	50° 06'N	4° 48'E
(h)	50° 54'N	7° 24'E	to	36° 22'N	85° 54'W

101. Odd one out

Which word is the odd one out?

CHAT, CHESS, CORNETIST, EON, PEST, RIDE, RIOTS, SINE, THUNDER, UNFIXED, ZONE

102. Where?

Where does CASTLE fit in the following list? *(Read left to right, top to bottom.)* See pages 145–8 for explanation.

WELL	CLEOPATRA	MILLER	IT	ANONYMOUS
RINGER	KID	BLUE	FIRE	LETTER
BARBARIAN	SALLY	EDWIN	BUFFET	FIGHT
FLASH	FAST	FOCUS	LEAVE	BOX
COMMON	BAG	MOUSTACHE	FORD	WAKE
ROBIN	WATER	BAPTIST	JOHN	DUSTER
PINK	GREEN	LOST	MEASURE	NOTHING
TREE	FLEA	JIM	A	BRUCE
DODGER	SORE	SAILOR	WOMAN	UNDER
TACK	SKETCH	SHOW	NIGHT	PRESSURE
DYNAMITE	POST	BARROW	HOLLY	WATCH

103. Cheltenham Tennis Festival

The Cheltenham Tennis Festival attracts some big names, and some odd ones. Games are best of three sets, and for this year the first-round results are below. A feature which was noticed in the scores continued through to the final. Who beat whom in the final, and what was the score?

First Round	Semi Finals	Final	Winner
Henman			
Gallek			
	Henman 6-4, 1-6, 6-1		
	Fastom 6-1, 2-6, 6-3		
Detlaf			
Fastom			
		? ?, ?, ?	
			? ?, ?, ?
		? ?, ?, ?	
Beckol			
Lytton			
	Lytton 6-4, 6-7, 6-1		
	Murray 4-6, 7-6, 6-1		
Murray			
Systan			

104. Bowling

I had a strange game of ten-pin bowling. I knocked down more pins than my opponent: indeed, my first bowl knocked down more than his, and my second, and my third, and so on for each bowl I sent down; yet I lost. What's the most I could have lost by?

105. Where am I?

My first is in Finland but not in Finnish

My second is in Estonia but not in Estonian

My third is in Albania but not in Albanian

My fourth is in Sweden but not in Swedish

My fifth is in Germany but not in German

My sixth is in Iceland but not in Icelandic

My last is in Wales but not in Welsh

My whole is in England but not in English

Where am I?

106. Complete the sequence

What word completes:

EMANATION, ALGA, MESSAGE, LYCHEE, SECRET, INCH, ARMLESS, ONSET, TEETER, RAPING, ?

107. Cities

Finished	'Town'	Mercer	Falcons	Crossing	Celtics	Anne Arundel
Richland	String	Wealthy	'Scotland'	Sir Walter	Bruins	Washington
Castle	Predators	Christopher	'Loaf'	Colts	Pollock	Dusty
Clift	Bracknell	President	Small	Warble	High	Powers
'Some'	James	Kings	Simon Templar	Bargains	Shawnee	Cleaner
Vehicles	Abraham	Broncos	Otto von	Cardin	Rubinstein	Dukakis
Ada	Laramie	Musical	Claus	Joaquin	Month	Opera

Which city is missing?

108. Word sequence IV

(a) ENOUGH, WORTHY, REHEAT, FURROW, VERIFY, EXISTS, ?

(b) INSECURE, WIZENED, REDISTRIBUTE, VERIFY, NUFFIELD, CHESSBOARD, BIENSEANCE, CHATTERBOX, ?

(c) CHAPLAINS, ABLATIVE, MAMMOGRAM, UNRELATED, RESPONSIBLE, ZEALOTRY, ELEVATE, THREATENS, RATIONED, ?

109. Drawing – lots

B was drawn by P

C was drawn by W D

C was drawn by L

D was drawn by S

J was drawn by P

N was drawn by S-H

P was drawn by B H

S was drawn by S H

V was drawn by D

110. Round Britain Quiz II

This style of question commemorates Round Britain Quiz *on BBC Radio 4: the format is one, long, cryptic question which has six parts to it, indicated by the letters a–f. To gain full marks you should identify all six parts. This will be sufficient to answer the question.*

A former UK Home Secretary known to the French as King John the fifteenth (a), Melchester Rovers' best known player (b) and a reference to Bosworth Field that has quite another purpose (c), each or all provide the end of a ribbed velvety cotton fabric (d), the city most associated with Heinrich Schliemann (e), and 2 of a set of 18 islands (f). How is this?

111. One word

What word links: David Harris-Jones, a Hindu goddess, handle with care, Hungarian rhapsodies, hire-purchase, Microsoft Windows, Cassius Clay, the first note of the tonic solfa scale and footwear?

112. Play detective

Associate words; divide into three groups; solve the mystery.

ALBERT	BRIDGE	COCKTAIL	COPYRIGHT	CORD
FINE	FITTED	MUTTART	OBELISK	O'HARA
PARTY	PLASTER	PLUMBING	PRIVATE	SIX-GUN
STRICTLY	THRONE	TOMATO		

113. Crossword

Complete the crossword. No clues have been provided for 1, 3, 4 and 6 down.

Across

2 Tetrahedral framework of silicon and oxygen in units of 1.137 litres, we hear (6)

4 Celebrity magazine featured in revolutionary novel by Chuck Palahniuk (5)

5 Little Scandinavian character belonging to me (2)

7 Black marketeer appears as notable individuals return (4)

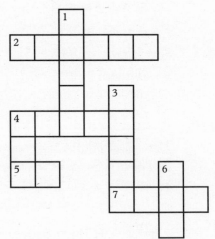

114. What is the highest?

In English there are 42 of which the highest is 88. In German there are only 12 and the highest is 80. In French there are a lot more. What is the highest?

115. Solve ALL clues

Solve all clues. The number of letters in the answer to each clue is the same as the length of the first word in that clue.

Nighties change direction in the Thatcher era
Author does at end of chapter
Fight following article, roughly
Hole-dwelling blemish
Auden initially, before questioning a location
Denmark, say, put right into Devon, say
Starts, for example, in containers for rubbish

116. Name dropping IV

The following words used to contain some famous names, before they became stars.

For example: *et cho*ol = Mark Lester (MARKet choLESTERol)

(a) c*bis af*   (TV)

(b) bi*ar des*   (dictator)

(c) *ot f*e   (dictator)

(d) es*o be*r   (music)

(e) ch* *ate   (music)

(f) ro*ud *rce (sport)

(g) b*ce pos*   (scientist)

(h) ca*r ch*ge (comedy)

(i) la*or s*le   (actor)

(j) p******   (1960s TV actor)

117. Sums

$$\frac{\text{Ways to leave your lover} \times \text{Red balloons}}{\text{Tribes}} - \text{Seas of Rhye} = ?$$

118. Whose parcel?

The parcel arrived exactly on time, a week before Christmas. The paper was patterned with squares forming a simple grid, each square containing a single letter. The letters in each row spelled out a brief repeating message, and every row contained the same message. One column contained the letters ITYAYRHH repeated several times; the next-but-one column along contained the letters TDAHIHAA similarly repeated. Who was the recipient of the parcel?

119. Word sequence V

What word could follow:

(a) FUN, ABLE, OUR, BIER, BRAVE, COIN, RUDE, SEAT, ?

(b) NIL, DEMO, DAY, NAME, SPILL, UP, READ, ?

120. Properties V

The words to the left of the colon all have a (non-trivial) property that the word to the right of the colon doesn't have. What is that property?

(a) AGE, BREATH, DOUBT, END, FACE, GROUND, HOPE: CLAY

(b) AS, END, ON, PER, PISH, PITY, RISE, ROOT: POUR

(c) BET, DIN, DIODE, GRATE, MINOR, ONION, PEAT, SLIM, TIN, URN: QUEUE

(d) BLADDER, CANDLE, EARTH, GINGER, IVORY, LOCK, MONKEY, PIG, SLEEVE, WING: FALSE

121. Underground

What follows:

MORDEN, MILE END, MOORGATE, MONUMENT, MAIDA VALE, MANOR HOUSE, MOOR PARK, MANSION HOUSE, MARYLEBONE, MILL HILL EAST, MARBLE ARCH, ?

122. Character actors

(a) Johnny Depp ____ Jay Silverheels

(b) Matt Smith ____ ____ Winston Smith

(c) Pope Pius XII ____ ____ ____ Queen Elizabeth II

(d) Ewan McGregor ____ ____ ____ ____ Commissioner Gordon

(e) Steve McQueen ____ ____ ____ ____ ____ Kevin Costner

(f) Denzel Washington ____ ____ ____ ____ ____ ____ Betelgeuse

123. Apposite phrase

Solve, giving the apposite phrase. Spaces are omitted.

(a) RESURRECTION, UNFINISHED, WAGNER, ANTARCTICA, LITTLE, 1917, ORGAN, ITALIAN, REFORMATION, SPRING

(b) 1553, 1760, 1216, 1461, 1189, 1625, 1660, 1413: 1558, 1216, 1714, 1830, 1689, 1216, 1547

(c) 1364, 1380, 1422, 922, 877, 1270, 1422, 1515, 1589, 893, 1574, 481, 893, 936, 986

(d) 1/12, 1/8, 7/1, 1/9, 4/10, 1/2, 4/9, 5/3, 7/11, 4/7, 4/6, 2/5, 3/3

(e) 1, 3, 2, 15, 3, 21, 4, 14, 5, 20, 6, 4, 7, 18, 8, 1, 9, 3, 10, 21, 11, 12, 12, 1

124. Complete the table

6	?	F
10	The Virgin	R
14	?	F
?	Pumping	R
29	?	R
?	Monster	R
33	?	?
?	Streak	F
78	?	?
79	?	F
80	Rising	?
?	?	F
86	?	F

125. Number sequence IV

(a) 7, 8, 5, 5, 3, 4, 4, ?

(b) 2, 2, 2, 3, 2, 2, 4, 2, 3, ?

(c) 1, 2, 3, 2, 1, 2, 3, 4, 2, 1, 2, 3, 4, 3, 2, ?

126. Interpretation

O3 K8 S3 F8 W6 N2 V5 H7 W4 B3 I6 O5 K3 V6 C6 H6 P4

127. ... and in French?

In English it's (910, 911, 912). In German it's (511, 512, 513). What is it in French?

128. Word division

Divide the following 281 words into 26 sets. Each set contains between 7 and 15 words.

A×4 ACROSS AGAIN ALL×3 AM×2 AN×2 AND×4 ANGEL
ANOTHER×2 ANSWER AS×2 AT×2 AWAY BEGGING
BETTER×3 BEWARE BIRMINGHAM BURDEN BUT BY×2
CALL CAME CAN'T CAR CARE×2 CHALK CHILD CHORD
CLOSE CLUB COLD CONTEMPLATE 'CROSS DARED
DARK DAVID DID DO×4 DOES DON'T×2 DOOR DOWN×2
EASY EVIL'S FAR FATE FEEL×2 FOREIGN FUNNY GET
GETTING GO×3 GONNA HANGS HAVE HEAD HEARD
HEARTS HEAVEN HER×2 HERE HOME HOW×2 I×13 ICE
IF×2 I'M IMAGINE IN×6 IS×2 IT×5 IT'S×2 I'VE JOLENE×4
JUST×2 KNOCKING KNOW×4 LIES LIFE LIGHT LIKE
LONDON LOOK LORD LOVE×2 LOWLY LURKING MAN
ME×2 MELTING MET MIDNIGHT MUCH×2 MY×7 MYSELF
NAPOLEON NEIGHBORHOOD NEVER×3 NIGHT 1965
NO×2 NOBODY OF×4 OLD ON×4 OR OUT PAIN PANIC×2
PASSED PLACE PLAYED PLAYGROUND PLEASE PLEASED
QUESTIONS RAINING RELAX REMEMBER×2 SAFEST
SAILING×2 SAME SAW SEA SECRET SEEMED×2 SIT
SLOWLY SO SOHO SOMETHING×2 SOMETIMES SOUNDS
STAND STRANGE STREETS×2 SURRENDER TAKE×2 TAKEN
TELL THAT×3 THE×12 THERE THERE'S×2 TO×5 TROUBLES
TRY VIETNAM WAIT WALL WANT WAR WAS WATER
WATERLOO WE×2 WELL WE'RE WHEN×2 WHERE WHO WHY
WINDOW WITH YESTERDAY YOU×10 YOU'D YOUR×2

129. Film sums

(a) THX – (Dalmatians × Weeks in a Balloon) = ?

(b) Missione Hydra = ?

(c) Acacia Avenue + (Charing Cross Road × Weddings and a Funeral) = ?

(d) A Space Odyssey + FBI Code – River Street = ?

(e) Catch × (Fahrenheit + Big Red) / Rifles = ?

130. ANTIDISESTABLISHMENTARIANS

The historian, aware of the sensitivity of the treatise he had just completed, decided to encipher it: but, knowing that a simple substitution offered little protection to such a long text, he came up with a scheme which he thought would defeat any would-be codebreaker. His enciphered text began HD LPPCLFXKFJXPKC LCGUN RFO CGNW KBFJSRWBYCKVF HA ISSYKDL EN . . . Some time later, he reached the last word of his text, only to find that it enciphered to itself – ANTIDISESTABLISHMENTARIANS!

(a) What was the plain text corresponding to LCGUN?

(b) What phrase did he use to generate the cipher alphabet?

131. Ode code

Here are some lines from Keats' 'Ode to Autumn':

SEASON OF MISTS AND MELLOW FRUITFULNESS,
CLOSE BOSOM-FRIEND OF THE MATURING SUN

(a) Explain how the first line of this enciphers as

LOSFLL WA WLLEE NUL LFNELF USRCISDSFOFT,

Here are some more lines (from different people), similarly enciphered. Decipher them.

(b) EE, IU TM V TTRS NEHIHGMLIEI MUARSM;

(c) ES OBNRA HOTAN ER DNM NDDA –

(d) EVSE JSX ITRTI?TW'W SU LES IEREI,

(e) EET NTLT TR TNRA TEOF LIN, EET ITET RG EEOM ENRN,

(f) GA LSR B ILADYE DDAI NYT DO PWELS,

(g) THH, HOLD! ITTI YGUHB RNRWIRI EKKIDH AKTREL EDTKSE?

132. **Which?**

The following list of 55 words can be divided into 10 sets, all of different lengths. Put another way, there is one set of 10 words, one set of 9 words, one set of 8 words, etc. What is the word in the set of one? See pages 148–9 for explanation.

ALL	AUGUST	BABY	BAR	BERRY
BLACK	BORODIN	BRENT	BRITTEN	BUTENE
CANADA	CHESTNUT	COUP	ELEVATES	FINE
FOR	FORGOTTEN	GOLDEN	GRAND	GREAT
KESH	LACROSSE	LED	LIECHTENSTEIN	LISTEN
MAJESTIC	MICKEY	MOTHER	MOUNTBATTEN	NOBLE
OMNICOMPETENCE	ONE	OPERATIC	PALE	PORTUGAL
PRAYER	RARE	RED	SIGHT	SNOW
SOMBRERO	STENCH	SUPERB	SUPERINTENDANT	TENNIS
THE	THERMOMETER	TIRAMISU	TITANIA	TRIDENT
UNDERLAP	V	WHITE	WOMEN	ZEUS

133. **Identify the following**

(a) London football team

(b) Boundary between military powers

(c) Soviet intelligence

(d) Swindon band

(e) Radio band

(f) Boy band

(g) Facility in Aldermaston

(h) Promissory note

(i) What city is left over?

134. **Odd one out**

Identify the odd one out.

(a) ACCEDE, BAMBOO, BANTAM, BARLEY, CURDLE, DAMSEL, DECODE, DEMURE, FIESTA, FIGURE

(b) BEHIND, CYMBAL, EXPIRY, FROZEN, INJECT, POWDER, SQUASH, UPWARD, WINTER, ZITHER

135. In the archive again

Mooching about in the ever expanding Kwiz archive we came across a table of letters. In the table 4 pairs of letters were highlighted (as shown below). Three of these were notes from the tonic sol-fa scale – DO, RE and FA. The 4th was a Greek letter (shown as ?? below). Which letter?

			?	
			?	
	F	A		
R	E	D		
		O		

136. Where?

Where does YULE fit in the following list? *(Read left to right, top to bottom.)* See pages 145–8 for explanation.

LAST	FITTING	PEAT	MESSAGE	HORSE	LOVE
BELL	BUTCHER	PRINCE	SILVER	BULL	TOY
CALL	FOX	MA	TALL	TOM	OLD
HOTEL	GROUND	DOG	DONKEY	ROCK	CORPORAL
PETER	GAME	CHANCE	MAGICIAN'S	EGG	HOT
PARTRIDGE	PIG	RANGE	TOOTH	HAIR	PLATFORM
WEDDING	HOOK	BOBBY	MINT	MINDS	FISH
STORM	VOYAGE	LION	DOWN	HEART	KING

137. Word sequence VI

(a) MILDRED, VIOLA, ELLEN, MONA, JOCELYN, SHEILA, ULRICA, ?

(b) OTHER, UTTER, PITFALL, CHUFFING, PROOFS, CONCESSION, SURPRISED, COMPREHEND, ?

(c) FREE, CHICK, TOWN, BOX, DUMPS, CLOVER, QUAY, CRAZY, ?

138. Where?

Where does ANIMAL fit in the following list? *(Read left to right, top to bottom.)* See pages 145–8 for explanation.

WEASEL	WOLF	ANTELOPE	BISON	DEER
CAMEL	HERON	SWAN	OSTRICH	TERMITE
TOAD	RHINOCEROS	TORTOISE	ALPACA	RAT
COYOTE	DUCK	SQUIRREL	EEL	SHEEP
HAWK	EAGLE	HARE	GOOSE	VULTURE
LOBSTER	POLECAT	OTTER	HEDGEHOG	ASS
KANGAROO	RABBIT	CROCODILE	SALMON	LEOPARD
BEAVER	PENGUIN	HORSE	MOUSE	CROW
HORNET	OWL	WHALE	LION	WALRUS
REINDEER	SHARK	SEAL	CHIMPANZEE	BADGER
MONKEY	BEAR	PIGEON	CAT	FOX

139. Kwestion I

Solve the following: 'In this sentence there are six. This kwestion contains twenty-two. How many are to be found in this sentence?'

140. A to Z

For each letter, we haven't provided the full list, but a selection of items, given in the correct order.

A DQ, KMKY, MMM

B IAS4U, O, INAGNYAW, ILR'N'R

C TBIB, YSS, DLTSGDOM

D OYH, TSOTC, LA(IBIL), DYWM

E SNA, IJDKWTDWM, THBTB

F F, FF, IWYS(P1&2)

G TBITB, G, DOTSOHS, YCCMA

H H, SOTSA, V-2S, TSLOA

I I, JG, GST, HDYS?

J P, LY, C, M

K NS, DI, NYT, M

L NWNC, CYBL, BS, RS, E

M ES, SBT, NEM, OWAM

N SLTS, IB, CAYA, L

O OPO, OPT, OH

P G&B, P, TTE, TIAL

Q TRM, IO, BB, LROM

R T, ER, HTAE, YS, TNK

S W, SYBT, 2B1, WDYTYA

T WBSS, T, BI, BJ

U CWTAE, STCFTHOTS, SSOSFAGTIACAGWAP

V WIME, PJ, ETS, POT

W SBS, NYD, THBAO

X 2H, IMA, TO, AIS, W

Y YSMR(LAR), ITD, LCBTM

Z E, D, Z, W, Z, B-BC

141. Kwestion II

Solve the following: 'In this sentence there are nine. This kwestion contains twenty-one. How many are to be found in this sentence?'

142. Sequences

What word might fit in the following sequences:

(a) OOLITHS, ?, ASSURES, PROOFED, DISLIKE, SLEEVES, EXISTED, ISOLATE, KETCHUP, EVENING, BLIGHTY

(b) HUNDRED, ?, PLEBIAN, CONCERT, AFFABLE, ANEMONE, MUSTARD, GALILEE, INSECTS, CHIPSET

(c) OFFICER, SUMMARY, ICEFLOE, CAHOOTS, BETIDES, COMFORT, ? PLINTHS, BAYONET

And with a slight difference:

(d) ANKLES, BUNKER, ?, WITHER, WEIGHS, OUTAGE

Which word listed in part (b) above would fit in:

(e) HEARTHS, SENORAS, TWINSET, ?, ACHTUNG

Which number would fit in:

(f) 016, ?, 484, 529, 676, 289

143. Christmas movies

(a) I A OEU IE

(b) IAE O I-OU EE

(c) OIA I

(d) EU O E I OE IA

(e) EO OOO

(f) E UE IA AO

(g) OE AUA

(h) E IAE EOE IA

(i) E OA

(j) IE A E A

(k) E OA EE

(l) E OU IE A

144. Kwestion III

Solve the following: 'In this sentence there are three. This kwestion contains eight. How many are to be found in this sentence?'

145. What connects?

Stewart Chail, Richard Heskit, Will Manby, Janis Borne, Jake McRigg

146. Which number?

The answer to each of the 4 parts of this question consists of just 2 letters. When you have solved each part, write the answers above each other in order to form a 4×2 grid of letters. With what 4-digit number can this grid be associated?

The words to the left of the colon all have a (non-trivial) property that the word to the right of the colon doesn't have. What is that property?

(a) BIN, BOWLS, LATE, LIE, RAINED: EXAMPLE

What word follows:

(b) WITHOUT, SATNAV, UPPER, QUIT, SAP, ?

What connects:

(c) GUSTED, MASTICATE, EIGHT, AWARE, MODERN, STITCH, RAGOUT, HURLED?

(d) BAD FEELING, TARANTINO FILM, LONDON THOROUGHFARE, CONFUSED, MUSTER?

147. Literary links

There follows a list of the surnames of various writers, then a list of numbers. Match each name with one of the numbers. What name could be associated with the remaining number?

BATES, BURROUGHS, CLARKE, GARDNER, HEINLEIN, JAMES, LEWIS, SHELLEY, WELLS, WOOLF

2, 23, 33, 46, 55, 68, 80, 82, 88, 89, 99

148. Squares

What property do the squares 81, 100 and 576 have, which no other square does?

149. Exception

The following words can be divided into two sets, with one exception. Which word is the exception?

ARTFORM, AVERAGE, BIRD, CLOWN, DITCH, DOG, EXECUTIONER, FATHER, FORBIDDANCE, GOODBYE, GRANDMOTHER, MOTHER, SENILE, STUNNER, TOY

150. Properties VI

The words to the left of the colon all have a (non-trivial) property that the word to the right of the colon doesn't have. What is that property?

(a) CALL, CHORE, HYMN, OMEN, RAN, RAT: AMUSE

(b) ASK, EGO, HIRE, LAW, SOUR, TUCK: NIT

151. Which?

The following list of 55 words can be divided into 10 sets, all of different lengths. Put another way, there is one set of 10 words, one set of 9 words, one set of 8 words, etc. What is the word in the set of one? See pages 148–9 for explanation.

AKBAR	AN	ANTHONY	ATLANTIC	AUTHOR
BAIT	BIAS	BOLEYN	BUSH	CALENDAR
CHANCELLOR	CHARACTERS	CHEER	CHENEY	COLIN
COLOSSUS	DINNER	EYOT	FETE	FREIGHT
GOOD	GREEN	IN	JACK	JENNINGS
JIM	KAYTE	KHAN	KING	KNOWS
LAND	LATE	LAW	LENIN	MAYOR
OF	OPENING	OVERT	PENINSULAR	PFENNIG
PRIDE	SEARCH	SCOTT	SHIP	SINGH
SINNED	SISTER	SMALL	SPINNEY	STRAIGHT
TENNIS	THWAITE	TIME	TIPTOES	WOOD

152. Follow the directions

Find the odd 2 out. Also, one entry in the list is worth two others combined – explain this.

08° 42'N   05° 58'W,   30° 10'S   23° 08'E,   32° 04'N   100° 41'W,

32° 16'N   98° 50'W,   34° 33'N   96° 59'W,   35° 25'N   114° 11'W,

36° 40'N   115° 59'W,   41° 58'N   02° 47'E,   44° 21'N   103° 46'W,

47° 24'N   79° 41'W,   58° 59'N   161° 48'W,   61° 57'N   128° 12'W

153. Number sequence V

(a) 8, 7, 1, 3, 3, 12, 13, 17, 21, ?

(b) 9, 1, 9, 20, 7, 11, 15, 13, 17, ?

(c) 1, 2, 4, 5, 6, 8, 10, 40, 46, 60, 61, 64, 80, 84, ?

154. Banned bands

A number of bands have been banned from the following radio playlist (e.g. pu*le *ic = puZZle TOPic = ZZ Top):

(a) ad*ced napht*e   (g) fi*an

(b) hon* en*ce   (h) b*g s*ler

(c) e*ore   (i) si*y b*s

(d) flam*d *e   (j) e*y *al

(e) c*ge   (k) e** no*

(f) e*h ma*ap   (l) re*e 3.1* . . .

155. Location, location, location

Put the following items into order:

13° 24' N 98° 31' E,   18° 20' N 74° 22' E,   32° 10' N 48° 15' E,

37° 48' N 01° 34'W,   41° 31' N 00° 21' E,   50° 40' N 03° 27'W,

50° 57' N 01° 51' E,   53° 34' N 08° 16' E,   55° 01' N 08° 26' E

156. Wordbox

A wordbox is a rectangle of letters in which a set of words can be read by starting at one letter and moving to an adjacent letter in a row or column, not diagonally, and not staying still. Retracing steps and reusing letters is allowed. Thus a wordbox might be:

```
D    O    R    F

I    G    E    S

T    A    O    T
```

which contains DODO, DOG, FROG, STOAT, TIGER and TIT, but doesn't contain GOAT or GEESE. This wordbox is of size 12, i.e. 3 (rows) × 4 (columns).

What is the smallest wordbox which contains the names of all the seven dwarfs: BASHFUL, DOC, DOPEY, GRUMPY, HAPPY, SLEEPY, SNEEZY?

157. Where?

Where does SOFT fit in the following list? *(Read left to right, top to bottom.)* See pages 145–8 for explanation.

BEND	PUDDING	EARLY	ROLL	WITCH
MAN	SPIN	WINDSOR	BABY	DOCK
DOG	CROSS	HAND	WEIGHT	FLAT
UP	BAG	POSTAL	EVEN	PIG
ROW	HEAD	TIMES	WICKET	KNOCK
SQUARE	BOW	WATER	GENERAL	MONEY
THIRD	MIND	DAILY	REGIMENT	WATCH
POLE	HARD	MID	RUN	VOLCANIC
MOUTH	COVER	PLAIN	GRANGE	SIGHT
APRIL	LOCK	DIVER	MUSIC	NO
SPINAL	BOOT	ABROAD	NECK	SISTERS

158. Which?

The following list of 55 words can be divided into 10 sets, all of different lengths. Put another way, there is one set of 10 words, one set of 9 words, one set of 8 words, etc. What is the word in the set of one? See pages 148–9 for explanation.

BOXER	CALMNESS	CARMEN	CHAMBER	CLIP
ENGINEER	ERASE	ETA	EUTROPY	EXTRA
FINEST	FOXTROT	FLOURISH	FOREMOST	FREEDOMS
GEORGE	GOLF	HARRIS	HORSEMEN	IMPEL
IRON	JEROME	JULY	JUXTAPOSE	LEADING
MARIAN	MARYLAND	MELBOURNE	NEXT	NITROGEN
OLD	OPTIMAL	OVER	OXEN	PARIS
PARLOUR	POLITICIAN	POSTER	SAXOPHONE	SEASONS
SINE	SPEECH	STRUGGLE	SUPREME	SUSSEX
TEN	TERRITORIAL	TEXAS	THAMES	UNSURPASSED
URALIC	VIOLET	VITAL	VOYAGE	WAXY

159. Telephone numbers

In the UK telephone numbers have 11 digits, starting with a nought. Some of these numbers are personalized, perhaps incorporating the first digit, perhaps not. Whose numbers might these belong to (if they happened to live in the UK, and in some cases were still alive!)?

(a) 01123 581 321

(b) 04931 151 419

(c) 04367 432 874

(d) 05660 787 718

(e) 01198 149 128

160. Girl's name

Which 6-letter girl's name could begin the following list?

?, HAD, INLAND, RAN, MAN, PAIN

161. First and last

The following sets contain the first and last letters of members
of groups. For example if the group was FLOPSY, MOPSY,
COTTONTAIL, PETER, the set of letters would be CFLMPRYY.
What are the groups?

(a) BCDDGHLSSYYYYY

(b) EJKLMMNW

(c) AAACCEEEEEEMOOPTTU

(d) ABFFGGILMMNNOPRSSY

(e) AADNPSSS

162. Sum clue

Miller's lover and bed artist on American naval vessel number nine
(4,5,3) = ?

163. Concerned amateur

An amateur cryptographer and scientist enciphered a 3-word phrase
by using a key consisting of the initial letters of the phrase, repeated as
required. What is the phrase?

XIUSM KMLUAV FATOULHD

164. Round Britain Quiz III

This style of question commemorates Round Britain Quiz *on BBC Radio
4: the format is one, long, cryptic question which has six parts to it, indicated
by the letters a–f. To gain full marks you should identify all six parts. This
will be sufficient to answer the question.*

If the originators of Dupin (a), 'he would, wouldn't he' (b) and
prandial nudity (c) were partially confused, they would lead to
someone (d) whose most famous creation (e) was later connected to a
world record for the 100 metres, set by (f). How is this?

165. Bizarre Pokémon

If (JB,2005) + (JB,2005) = (IM,1978), complete the following:

  (a) (JB,19??) = ?
  (b) (????, ????) = ?????? ??? ??????

166. Language

Explain the following: falkun, homem de guerra, kock, lavoro, moed, mosca, raccordement, schäfer, stobhach, tre, zevk

167. Divide into pairs

BICKER, BUSK, CAMEL, CURRY, GAUNTLET, MALLET, MARTYR, QUEEN, STALLING, WORTHY

168. What's the point?

Identify the twelve items in the following list:

AAAₐₐAₐₐAₐₐAᴄCCCᴄᴄCᴅDᴅᴇEEEₑₑₑEₑₑғғGGₒₕHₕₗᵢᵢᵢⱼKLLLLₗₘMMₘₘNNₙNₙₙₙₙₙₙNₙₙₒOOOₚᵣRRᵣRₛSSSSₛₛₛₜₜₜTₜₜTY

169. Shapes

Reorganize these shapes to make an appropriate picture. Some may need to be rotated, but none needs to be reflected or scaled in size.

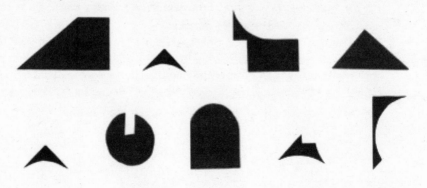

Colourful characters

Identify these people.

(a)

(b)

(c)

(d)

(e)

(f)

(g)

(h)

(i)

(j)

(k)

(l)

(m)

(n)

(o)

Next in sequence

Identify and divide into groups

a.

b.

c.

d.

e.

f.

g.

h.

i.

j.

k.

l.

m.

n.

o.

p.

q.

r.

s.

t.

u.

v.

w.

x.

y.

z.

Identify the people

Identify the people pictured below and divide into inappropriately matched couples.

a. b. c. d. e. f.

g. h. i. j. k. l. m.

n. o. p. q. r. s.

t. u. v. w. x. y. z.

Pictures

Which is the odd one out?

Celebrity Sudoku

	1	2	3	4	5	6	7	8	9
A									
B									
C									
D									
E									
F									
G									
H									
I									

i) Identify the people in the grid above

ii) Identify the people on the right and give the grid reference of the squares they go in

a.   b.   c.   d.   e.

f.   g.   h.   i.   j.

Townsfolk

Identify these people.

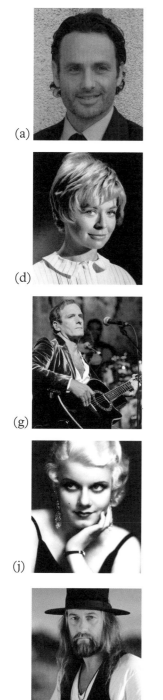
(a)

(b)

(c)

(d)

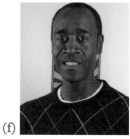
(e)

(f)

(g)

(h)

(i)

(j)

(k)

(l)

(m)

(n)

(o)

Capital cities

Identify these people and things.

(a)

(b)

(c)

(d)

(e)

(f)

(g)

(h)

(i)

(j)

(k)

(l)

Fill in the gaps

14   24   32   ?   ?

Sums

(a)

= ?

(b)

= ?

Solutions to these puzzles can be found in the Answers section at
the back of the book, on page 320.

This cryptic crossword was first published as a competition in the *Daily Telegraph* in January 1942. It is thought to have been used to recruit code breakers to Bletchley Park during World War II.

ACROSS

1. A stage company (6)
4. The direct route preferred by the Roundheads (two words—6, 3)
9. One of the evergreens (6)
10. Scented (8)
12. Course with an apt finish (5)
13. Much that could be got from a timber merchant (two words—5, 4)
15. We have nothing and are in debt (3)
16. Pretend (5)
17. Is this town ready for a flood? (6)
22. The little fellow has some beer: it makes me lose color, I say (6)
24. Fashion of a famous French family (5)
27. Tree (3)
28. One might of course use this tool to core an apple (9)
31. Once used for unofficial currency (5)
32. Those well brought up help these over stiles (two words—4, 4)
33. A sport in a hurry (6)
34. Is the workshop that produces this part of a motor a hush-hush affair? (8)
35. An illumination functioning (6)

DOWN

1. Official instruction not to forget the servants (8)
2. Said to be a remedy for a burn (two words—5,3)
3. Kind of alias (9)
5. A disagreeable company (5)
6. Debtors may have to this money for their debts unless of course their creditors do it to the debts (5)
7. Boat that should be able to suit everyone (6)
8. Gear (6)
11. Business with the end in sight (6)
14. The right sort of woman to start a dame school (3)
18. "The War" (anag) (6)
19. When hammering take care to hit this (two words—5, 4)
20. Making a sound as a bell (8)
21. Half a fortnight of old (8)
23. Bird, dish of coin (3)
25. This sign of the zodiac has no connection with the Fishes (6)
26. A preservative of the teeth (6)
29. Famous sculptor (5)
30. This part of the locomotive engine could sound familiar to the golfer (5)

Answers to the crossword can be found at www.gchqpuzzlebook.co.uk

170. Wordgrid

Reconstruct the 16×16 grid and work out which one of the 13 is missing.

A	D	G	E
D	A	C	A
A	C	H	A
A	A	C	A

N	S	A	D
C	E	A	N
A	C	A	N
A	C	A	N

L	A	C	A
R	O	T	C
N	A	T	D
A	C	A	T

N	A	C	A
A	C	A	N
A	A	N	A
Y	C	A	N

E	C	N	A
B	T	T	A
E	I	O	D
U	G	D	W

T	N	T	N
I	V	O	A
C	T	R	A
C	N	O	A

Q	A	I	D
N	A	D	N
A	D	A	C
O	L	L	E

A	N	A	D
A	C	A	I
A	C	F	A
N	A	A	D

X	C	A	N
A	C	A	N
A	D	A	C
A	D	A	C

C	A	N	L
A	C	A	A
C	Q	A	N
I	N	K	W

P	I	N	N
N	S	A	D
N	A	T	D
Y	N	A	J

C	C	A	N
A	H	N	A
N	A	A	D
N	A	D	R

A	D	A	E
N	A	R	D
A	F	D	A
D	A	E	F

I	W	A	C
A	H	C	A
A	I	C	A
D	T	A	C

O	E	A	D
N	H	A	D
C	O	N	A
D	R	A	S

A	D	A	I
A	D	R	R
A	E	N	A
D	A	N	A

171. Transformations

(a) How does:

S become H	C become I	B become J	L become K
B become M	K become O	E become T	

(b) How does:

NF become C	FS become D	A become E	C become G
B become L	A become Y	R become Z	UV become BF

172. Letter sequence III

(a) A, T, G, C, L, V, L, S, S, C, A, ?

(b) A, E, A, P, A, U, U, U, E, C, O, ?

(c) E, O, R, U, V, X, V, G, N, ?

(d) C, U, T, S, I, U, N, ?

(e) N, S, U, O, K, E, S, E, N, I, O, I, ?

173. Chains

(a) GOLDEN ____ SUGAR

(b) TRUE ____ ____ RIVER

(c) WONDERFUL ____ ____ ____ MACHINE

(d) BLACKBERRY ____ ____ ____ ____ PRESSURE

(e) MA ____ ____ ____ ____ ____ BRIGADE

174. Word sequence VII

(a) EMPTY, CELLS, THEFT, STREW, ?

(b) CROW, ROBIN, OSTRICH, SPARROW, PEEWIT, EAGLE, ALBATROSS, ?

(c) COMA, BATCH, DANCER, WARRANTED, SCARE, INFEST, GARDEN, ?

(d) SLAB, CRIB, CARD, SLED, LEAF, GOLF, HUNG, DISH, JILT, JACK, ?

(e) JACKAL, MILKED, MEANLY, YEOMAN, SPOONS, ?

175. What comes next?

(a) B, SB, M, C, Q, SQ, ?, ?

(b) O, C^4, O^2, C^2, O, C, O^3, C^6, O, C, ?, ...

(c) TF, F, T, S, W, ?

176. Strictly Come Dancing

Strictly Come Dancing, also known as *Dancing with the Stars*, is a television dancing competition. Professional dancers are paired with celebrities (of different sex), and each week they are marked by a set of judges, and then by an audience phone vote. The lowest scoring couple overall is eliminated from the competition. At one stage in a recent run there were just five couples left, and it was at the stage in the programme that only the judges had marked them. Each couple had got a different score.

Anna was pleased with second place as it made her the highest placed female professional dancer. However, Greg was disappointed with his position, particularly as Nigel and his professional partner had beaten him. The model burst into tears because the reporter finished higher up the table. Teresa was delighted to have done so well, as was her professional partner, especially as he finished ahead of his wife Lilia (interestingly the five professional dancers included two married couples). I noticed a couple of other things too – the winning celebrity had the same name as the fourth-placed professional dancer, who partnered the businessman; and that if you read in order from highest to lowest score, the initial letters of the names of the professional dancers, the initial letters of the names of the celebrities, and the initial letters of the celebrities' real professions, all spelled words.

Suggest a name for the last-placed professional dancer (his name does not appear above).

177. What's the film?

HAND, LONE, TWO, WHOLE; LADY, LIFE, SUCH, WORKMAN; BAND, DIG, ORB, PULP; EARS, GUNS, MUGS, SNAPS.

178. Lists

The following list is in order. Explain why.

OM, JT, UM, VT, SA, IS, BE, TB, OI, MG

179. Alphabetic groupings

Explain the following alphabetic groupings.

(a) EGKLMPXZ / ABDHIORT / JQSVY / CFUW / N

(b) ET / AIMN / DGKORSUW / BCFHJLPQVXYZ

(c) A / BCEIK / DFHJLMOSU / GNPRTVWXZ / QY

(d) BCEGKMQSW / ADFHIJLNOPRTUVXYZ

180. Where?

Where does CHANCE fit in the following list? *(Read left to right, top to bottom.)* See pages 145–8 for explanation.

ACID	HARD	HER	WEDDED	CLEAN
SPROUT	ADA	BOY	GEORGE	ON
PIN	MARK	GONE	PARTY	OCEAN'S
LATE	SHOW	FIND	TOP	LEVEL
PROFESSIONAL	GLOVE	ANY	PAMPAS	GIVE
USE	CROSS	BANE	GARY	LAND
WALK	PULL	BEAN	PRIDE	ENVELOPE
NEAR	LOCH	LANDS	JOE	WINDOWS
WAY	FORWARD	PICK	FLOWER	SPRING
ROUTE	IN	SYNDROME	TIMES	NUMBER
TOUCH	CABER	CATCH	WOODS	BOOT

181. Next in sequence

In, The, Paul, Paul, Paul, Paul, Paul, Paul, Paul, Paul, Paul, Paul, Paul, Paul, Paul, Paul, ?

182. What connects?

(a) 5°39'N 169°08'E   14°02'N 0°03'W   25°00'N 55°56'E
     43°15'N 2°55'W   52°49'N 18°19'E   66°13'N 72°15'E

(b) 10°30'S 39°01'E   10°23'N 9°19'W   18°02'S 65°27'W
    25°03'S 33°41'E   35°07'S 147°22'E   46°04'N 118°21'W

183. Which?

The following list of 55 words can be divided into 10 sets, all of different lengths. Put another way, there is one set of 10 words, one set of 9 words, one set of 8 words, etc. What is the word in the set of one? See pages 148–9 for explanation.

ABLE	AGE	ANT	ARCHBISHOP	BATTER
BELIEF	BLACK	CHEATING	CHRIS	CHUFF
CORAL	DEAD	DON	DRAWING	DRILLED
FIRING	FLAT	FORGOT	GLITCHES	GRAPEFRUIT
GRAPHITE	HAIR	HAS	HAT	HISTOGEN
IRINA	IRISH	KING	KNIGHTED	KNOWN
LARRY	MAGNIFICENT	MEGALITH	MIDWIFE	NORTH
NOTED	NOTORIOUS	OR	OUTWEIGH	QUALIFIED
READING	RECOGNIZED	RED	RENNES	RENOWNED
ROLLING	ROT	SAFETY	SKEWER	SKIRT
TAILS	THEURGIC	TIE	TRANSFER	TURN

184. Series sums

What is the next number in the following series, each of which has been formed by adding 2 other series together:

(a) 2, 6, 13, 24, 41, 68, ?

(b) 3, 4, 7, 10, 16, 21, 30, 40, ?

(c) 5, 8, 5, 9, 7, 17, 3, 14, ?

(d) 4, 5, 8, 8, 9, 9, 12, 13, 13, 13, ?

(e) 6, 8, 9, 10, 8, 8, 10, 8, 8, 7, 7, 10, ?

(f) 4, 8, 11, 11, 15, 16, 15, 17, 13, 13, 20, 24, 24, 25, 28, 30, ?

185. Which animal?

A town in southern England has something in common with two cities in Italy, one in Greece, one in Lebanon and an animal. Which animal?

186. Coldplay V

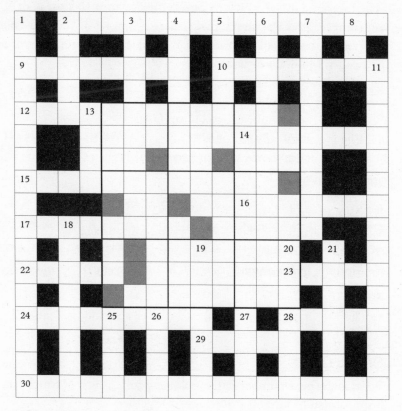

Across

2 Larry at the seaside? (5,2,1,7)
9 Levitate sulphur: two gallons is a minute amount (8)
10 Controls voltage as the rot mutates (8)
12 Abhors confusion, makes my little pony (6)
14 Most depressed dad sets puzzle (7)
15 Look at squashed roses – they're ugly (8)
16 Stood beer for Kane mystery (7)
17 Stop him holding gouty nodules (5)
22 Writes music (5)
23 Gandhi's not right where she governed (5)
24 In hollow roll that is containing uranium confused with song bird (4,5)
28 Former political subversive group (left left) to be best (5)
29 I link cattle disease to northern playwright (5)
30 Avoid trouble with a handkerchief (4,4,4,5)

Down

1   Strange effect – thank boater in the wilderness (3,3,6,5)
2   Heavy seas? Not even in Hillingdon (5)
3   Ramp up with creole? Not in this time; in the mouth (8)
4   Caustic Capone with Hindu goddess (6)
5   Thomas author loses daughter, that's not right (4)
6   Tip: one hears chest muscles (4)
7   Is erred, odd, confused, mixed-up (10)
8   Egg is first of various avians (3)
11   Frantic alarm; noted title in shreds (15)
13   Listens to insects (4)
18   Unpoetic mess – keep it for later (3,2,3)
19   All the vowels without us left Ian to the winds (7)
20   After Ireland, Nicaragua begins promoting peace (7)
21   English feline caught by obligation. That'll teach him (7)
25   Nothing follows orchestra up in Scandanavia (4)
26   Editor is climbing face (4)
27   Takes advantage of purposes (4)

187.   Odd one out

In each of the following lists, all but one of the words share a strong common property. Which is the odd one out?

(a)   ABSURDITY, BACTERIUM, CUSTODIAN, DEFOLIANT, EQUITABLY, FLYWEIGHT, GERANIUMS, HESITANCY, INCUBATOR, JURYWOMAN

(b)   AUTHORISED, BIVOUACKED, COEQUALITY, DISCOURAGE, EXHAUSTION, FAVOURITES, GLAUCONITE, HOSPITABLE, INOCULATED, JOURNALISE

(c)   ADVENTURISM, CENTRIFUGAL, DISTURBANCE, DISGRACEFUL, FINGERHOLDS, GOLDFINCHES, IMPORTUNELY, OBTRUSIVELY, PALINDROMES, QUESTIONARY, SWITCHBLADE, UNSCRAMBLED

188. The killer was …

One of our colleagues, who is no longer with us, left the following clue to his killer:

SNLOU IAIRO SCSCN OLCIO IOIMM NVPAU OOORE CCCTN OISLP

Identify the killer.

189. Missing from the list

Which other US state should also be in the following list?

ASTATINE, ELEPHANT, EPSILON, ROMANIA, SUSANNAH, WALTER, WEST VIRGINIA

190. Number sequence VI

(a) 3, 8, 18, 9, 19, 20, 13, 1, ?

(b) 9, 14, 11, 8, 16, 7, 19, 3, 17, 2, 15, ?

191. Sums

IFOF – IFOG – IFOW + ILAG + WHFAL = ?

192. A to Z

A = EYAWTKAS*BWATA, AH, M, HAHS

B = BJ, ES, EW, POTA

C = TT, A, TL, T

D = C, TU, BOTV, M:I

E = PMFM, U, TBOMC, MR

F = TGOW, HGWMV, TS, TMWSLV

G = MPATHG, TB, B, TM

H = TLV, TMWKTM, NBN, P

I = HAD, ARWAV, ROTD, HE

J = BT, HC, LOTR, KK

K = POG, DSO:HILTSWALTB, ACO, TS

L = KFM, TBB, AAWIL, CTA

M = M, TLOTM, H, MV

N = M, I, BB, TP

O = TDC, TMTM, LSOH, TSW

P = C-D-S, RB, C, TP

Q = BBAC, TWOSW, SATSG, HTMYW

R = TIST, SBM, WHMS, M

S = J, ROTLA, JP, SL

T = RD, PF, JB, DP

U = T, CS, MJY, TAOPN

V = TR, BI, ST, HM

W = TSYI, WFTP, SLIH, TPLOSH

X = GSXDHH, ZHDGZ, QLSZS, BLY

Y = DN, FRWL, T, WUD

Z = BTTF, WFRR, FG, CA

193. How many U's

The square of letters below has had most of the letters replaced by dots. 7 letters, one each of N, O, P, Q, R, S, T are shown. How many U's are there in the completed square?

.	.	.	.	.	.
.	.	O	.	.	.
.	.	T	R	.	.
.	.	Q	S	.	.
.	.	P	.	.	.
.	.	N	.	.	.

194. Last word

What is the last word (unenciphered!) in the following sentence?

NO OO OO TON NO OO NOTN TR NOO OOB-BOOTRO

195. Film censorship

Identify the films that have been censored from the following list.

For example, ki*orie c*ot (1983) = ki**local**orie **chero**ot = LOCAL HERO.

(a) dispa* e*ient   (1980)

(b) para*logy   (1960)

(c) ar* af* ma*   (1964)

(d) po*e cli*   (1979)

(e) a*u orc*   (1988)

(f) s*e f*y   (1997)

(g) t*re   (1995)

(h) *ch c*ch   (2016)

(i) s*t   (1979)

(j) *o ba*d   (1960)

196. Features

In each of the following lists, a number of the words share a common feature, and the same number share a different, but related, feature. The remaining word shares neither feature – or in one case, both!

(a) BEGIN, OILED, ADEPT, POLKA, TRIED, BELOW, SOLID, TONIC, SPOKE, FIRST, GIPSY, WRONG, FORTY

(b) EERIE, CLASH, CHASM, SIEGE, NAVAL, TEPEE, SHANK, GLASS, PIPER, QUITE, WRITE

(c) FACED, CUBED, BEING, FIGHT, VAPID, ORGAN, LEMON, WRECK, FOCUS, EXTRA, SQUID, BURST, TORUS

(d) LADLE, CLEAN, CIGAR, BAILS, GAMES, HOARD, DAILY, AURAL, WINED, CANNY, SMILE, MANOR, LYRIC, WILES, BYTES

(e) LACED, ANGLE, LEASE, MINED, SPEED, ZEBRA, ELBOW, DEPOT, LEVER, ALLOY, MERIT, REBUT, VERSE

Which are the 'odd ones out'? What rule governs the ordering of the words?

197. Incomplete codeword

Some British newspapers contain a kind of crossword in which each non-blank square contains one of the numbers 1 through 26. Each number represents a different letter of the alphabet. Reading from top left to bottom right, the first letter that occurs is number 1, the next different letter that occurs is number 2 and so on. Kontestants will not be surprised to learn that the ubiquitous koffee kup has obliterated most of the first two rows . . .

1												
3	13	11	1	■	14	5	15	15	16	17	1	18
8	■	2	■	1	■	19	■	13	■	20	■	12
5	21	12	3	22	1	19	5	18	11	■	■	17
■	■	19	■	13	■	6	■	■	■	17	■	20
1	19	19	13	9	6	■	11	20	15	13	9	6
18	■	16	■	■	2	■	12	■	1	■	■	■
1	■	■	17	3	11	15	13	18	5	19	19	1
5	■	1	■	18	■	13	■	23	■	2	■	8
24	13	19	19	12	6	17	6	■	20	1	19	13
3	■	6	■	15	■	25	■	■	■	17	■	7
1	7	13	8	5	■	6	21	12	5	5	26	5

What are 1 across and 1 down?

198. Intercepted message

The following message was intercepted:

AFSDI VFTLJ FQXJB NXFHR XIHSX

The recipient obviously could not decrypt it, as the originator resent it as follows:

QXSTG VFXZB XILTB BNTFV VUHWB

(a) What did the message say?

(b) How was the key generated?

199. Christmas message

An unusual email was forwarded to the Kwiz office, which would seem to have a Christmas theme. It read:

> *Dear Julius,*
>
> *I have done what you suggested and it worked! We both re-engineered and reconstructed the thing, and all of our efforts were rewarded. But, just to see if I've been a fool and missed something vital, we spent this evening taking the sleigh to bits, and rebuilding it, yet again. I never thought it would take so long, but the night ended happily. Hoorah!*
>
> *With much gratitude*
> *Santa Claus*

Our correspondent felt that Santa had indeed missed something – but we were able to show that he hadn't.

(a) What did our correspondent think had been missed?

(b) How did we reassure him?

200. Properties

Each of the following lists contains a number of words which have a strong common property, and one which has a similar, but distinctively different property. Which are the odd ones out?

(a) AVENUE, BINDER, COURAGE, FORTUNE, LEMUR, MERCHANT, NONPLUS, TRESPASSER

(b) CLAPPER, CLUMPS, DELUGING, DEPART, DOLMENS, HANDSOME, SMILED, STEADY, TELEGRAPHIC, TRACTORS

(c) DEMERGE, EMANATE, GRATIN, HUNTERS, MALAR, OVERS, RAMBLE, SPOTTER, TRIO, WANE

(d) APPEARS, COMATOSE, ECLIPSE, EFFETE, JEOPARDISE, NECROMANCY, PURISTS, RICHEST, URCHIN, VENOSE

(e) ALLIANCE, BARGAINED, CROUPIER, ENGAGES, PRICKLE, PROTOPLASM, SHANTIES, SITUATION, STINGING, THANKLESS

201. Where?

Where does FORTUNE fit into the following list? *(Read left to right, top to bottom.)* See pages 145–8 for explanation.

SEA	ARROWS	THEATRE	THAT	BAG
SIT	BEACH	WHITE	PUDDING	GIRLS
EAR	LIFT	BROWN	HOT	POTATO
SAFE	BEE	COURSE	STONE	CALIFORNIA
RUBBER	JONES	BIRD	FRIED	GRAM
IRON	BOX	BONE	FATS	CLOUD
BITING	RINGS	EMPIRE	DOC	SHOOTER
MINT	SEX	WHETHER	STATUS	WAY
RED	OR	NEVADA	ROLLING	PIGEON
DIRE	FOOTBALL	BACK	ARMS	WILLIAMS
MIND	SUFFER	CREEPER	IRVING	DOODLE

202. What error?

My three-year-old daughter and I were making up funny sentences. I said to her, 'The octopus wore a hat as she ate an apple by the volcano.' She said to me, 'A penguin lives in a cave, but a bee lives in a wigwam.' What mistake did my daughter make?

203. Complete the set

Bedazzlement, Highjack, Quantify, Unveiled, Yokewood, ?

204. The states

The following US states have a particular property:

ALABAMA, ALASKA, ARIZONA, ARKANSAS, CALIFORNIA, CONNECTICUT, MAINE, MICHIGAN, MINNESOTA, NEW JERSEY, NEW MEXICO, TENNESSEE, WASHINGTON, WEST VIRGINIA, WYOMING

RHODE ISLAND has the property twice. What is the property?

205. Missing member

The following is an encoded set of 26 entities. What is the missing item?

ADJIW	CQOE	DBSMPX	DVRKY	ECXCP	EGFSYZSF
FODUH	GSVO	HYYTER	ITSJLFQ	JAHROT	LIMWIGGIH
NGDBJ	NHSDHF	PELLANWNU	SMZZG	TRUMJAN	TXQBOCLOA
UCQRKCYRF	USVUOXXI	VDWENQC	???????	WLZTD	XQUFDUY
YVZREVPX	ZCBUTCFR				

206. I'm Sorry I Haven't a Clue

With acknowledgements (apologies) to *ISIHAC*, here's a round of 'one song encrypted by the key of another'. Each pair of songs is connected. What are the pair of songs?

(a) (K IPH'S LTS HJ) RPSKRBPISKJH; LISJR SR, CDIAB

(b) NIUHQU NIUHQU KU; CSY FUAS

(c) QNA PSLLAY VDQN QNA RLDGMA IG QIJ; KWU RTDOUD TPS KWU EAQOTP

(d) XGQ IFGU CX FACH; YGHQANL TYX QH KEN

207. Is it some education?

(a) Norman the Robot

(b) Avoid Guernsey beer

(c) Some wistful Earthmen

(d) Defrost hard blockage

(e) Thy diabolic fever

208. Word sequence VIII

(a) ME, ROW, MAIL, SWOON, JACKAL, PRONOUN, AMICABLE, PROTOZOON, ?

(b) WE, CITY, NOTICE, ELECTRON, PROVOCABLE, ?

(c) AMAZED, BRAYING, CRUXES, DROWNED, ELEVEN, FLOUTED, GRATES, HESSIAN, INURED, ?

(d) CHAIN, RIBBON, PRECIPICE, ADEQUATE, EAR, REFUSE, WEIGHTY, SHRUGGED, ACHIEVE, BEJEWELLED, PICKAXE, SMELLY, ?

209. Answer the question

74 1 85, 53 16, 8 10, 15 71 16, 28 10?

210. Which?

The following list of 55 words can be divided into 10 sets, all of different lengths. Put another way, there is one set of 10 words, one set of 9 words, one set of 8 words, etc. What is the word in the set of one? See pages 148–9 for explanation.

BACK	BERNOULLI	BLUE	BRONZE	BROWNE
CAP	CAR	COLIN	DARLING	DECK
DESTINY	DETEST	EDUCATION	ESPERANTO	ESTABLISH
EWER	FINAL	GLENDA	GOLDEN	HORS D'OEUVRES
IRON	JONES	KELLY	KLEIN	LINCOLN
MAGNETRON	MASTER	MIDDLE	MILK	OLD
PAL	PERCY	PERISH	POLLOCK	PORTAL
PROTOZOON	PUTTING	RECTANGLE	REVEREND	RILE
SAGE	SATISFY	SEA	SPACE	STONE
STRONTIUM	TALLY	TARDY	TAUT	TRANSPORT
TRYST	VALUE	VILLAGE	WATER	WINE

211. Be precise

Identify, precisely, the sets:

(a) BEKMQJ, EJP, LMPLQB, NCDJ, QEJJM, WLBCJH, YJCCBO

(b) ABHHZY, ASHHUY, EKUUY, JCN, JCUHY, MKAEOWS, PLWTUY

(c) AYLCVD, BISSRQCLRF, GLUGDAVD, GQSBIGUGVD, KGSGFCVD, SGUABCFRSCVD, SRPYFIKCD

212. Sums

(a) David Van Day + Thereza Bazar – Curtis Jackson = ?

(b) McKern + Portman + Cargill = ?

(c) (Fiedler × Sweeney) – Voskovec = ?

(d) ((Prince + Paul Hardcastle + Frankie Vaughan – Sea Level) × UB40) – The Bystanders = ?

213. Film censorship

A number of films have been censored from the following list. Identify them.

For example, ki*orie c*ot (1983) = kiLOCALorie cHEROot = LOCAL HERO.

(a) u*ia bur*dy   (1986)

(b) porc*ine   (2009)

(c) cu*d *hip   (1977)

(d) le*de pa*de   (1986)

(e) qu* st*   (1969)

(f) ca*e ca*e   (1962)

(g) p*r p*ic   (1977)

(h) a*s c*d   (1995)

(i) p*lu*   (1961)

(j) *ey 2.7*8 . . .   (1985)

214. 6 sets of 6

The following set of digits represents 36 entities, in alphabetical order:

1, 1, 6, 6, 5, 4, 5, 2, 2, 6, 6, 6, 3, 1, 5, 5, 1, 4, 6, 2, 2, 4, 5, 3, 3, 4, 3, 1, 2, 2, 4, 3, 5, 3, 1, 4

Two of the 2's are associated with all of the 6's. Explain!

215. Unalike words

Unalike words are pairs of words which have no repeated letters either within them or between them (e.g. UNALIKE and WORDS). Find the unalike words which maximize the product of their lengths.

216. Letters

Many letters of the alphabet can represent certain entities, and some of these entities can form 'sets' – for example, CFK can represent temperature scales (Centigrade/Celsius, Fahrenheit and Kelvin) and constitute a set of 3.

(a) Identify the set BCFHIKNOPSUVWY

(b) Identify the four sets – one of 7, two of 4 and one of 2 – in ACDEFIJKLMNPQSVWX

217. Odd word out

Another odd-word-out question: the themes in the following words are identical, but totally different! Which is the odd word out in each case?

(a) ANGLE, BRING, CLAMP, DIRTY, EXACT, FIELD, GRASS, HEART, IMAGE, JAUNT

(b) ABBEY, BURST, COURT, DRINK, ENJOY, FOUND, GIANT, HARMS, IDIOT, JUMPY

218. Fill in the gap

George	Holly	Michael	Geno
Benjamin	Andrew	Andrew	George
Rex	Don	Gordon	?

219. US state

(a) The following digraphs are a complete set. Why?

AL, AR, CA, CO, FL, GA, IN, LA, MD, MN, MO, MT, ND, NE, NH, PA, SC

(b) Which US state, associated with number (9), completes the set:

MONTANA (1), NEW HAMPSHIRE (1), OHIO (1), SOUTH CAROLINA (1), UTAH (1)

220. Which word is left?

From the list:

ALGAE, BULBS, COUCH, DEPTH, EASEL, FJORD, GENII, HOPED, JAMMY, KNIFE, LABEL, MAIMS, NIECE, ORGAN, PATHS, QUOTE, SEATS, TITHE, UPSET, VERVE, WASTE, WEAVE, WHIRL, WORTH, WRONG, YACHT

Eliminate 5 words which have a common property associated with their 1st letter;

then eliminate 5 words which have a common property associated with their 2nd letter;

then eliminate 5 words which have a common property associated with their 3rd letter;

then eliminate 5 words which have a common property associated with their 4th letter;

then eliminate 5 words which have a common property associated with their 5th letter.

Which word is left?

221. Number sequence VII

(a) 1, 2, 5, 13, 34, 89, 233, 610, 1597, ?

(b) 1, 1, 0, -1, 0, 7, 28, 79, 192, ?

(c) 1, 2, 2, 4, 2, 4, 2, 4, ?

(d) 0, 5, 2, 6, 3, 1, 5, 7, 8, 9, 4, ?

(e) ?, 30, 42, 54, 66, 78, 90, 144, 259

222. What comes next?

What is the next number in the series below? We will accept either of two possible answers. *(As a hint, finding the next-but-one number will certainly prove difficult, if not impossible.)*

6, 11, 23, 124, ?

223. A wordsearch fit for a knight

The grid below contains a number of words with a strong common theme, written into the grid using Knight moves. Thus, starting from the top-left, a possible word might be B-O-Y (Row 1 Col 1, Row 2 Col 3, Row 4 Col 2). Sadly, BOY is not an answer! Consecutive letters must come from different cells, but a cell can be used more than once in a word. Find as many words as you can.

B	H	A	Y	P	C	R	D	F	R
L	D	O	E	M	O	B	U	M	G
T	C	L	S	R	R	A	E	Y	O
U	Y	E	H	X	V	K	I	B	S
M	A	A	L	T	R	N	S	R	M
O	B	T	G	L	E	U	O	O	U
A	E	S	A	I	L	O	H	M	P
C	X	N	S	B	T	N	O	R	R
N	I	G	S	D	D	C	H	E	V
Y	E	A	O	K	O	T	A	G	I

224. Where?

Where does SYSTEM fit in the following list? *(Read left to right, top to bottom.)* See pages 145–8 for explanation.

RAT	UP	BILL	FOND	ROOT
FLY	BELL	BOY	FAITH	MITE
RECTAL	COMB	PLANT	MAN	POUR
WILD	PATE	TOM	TRAP	LAG
SEE	GRAND	MELON	BRIEF	HELL
GAME	ELBOW	SLENDER	COBWEB	SINGE
HERO	CHORE	FOUR	IRON	CHEAP
LEONINE	TREE	GIFT	MARE	ROT
SORE	SUREST	ROLL	LAP	LIVERS
HOOK	PEDANT	ANT	IRIS	EAR
BILLET	BACON	TOUR	BELCH	GUARD

225. Rugby

In a rugby union tournament a team gets 4 points for a win, 0 for a loss – plus 1 bonus point for scoring 4 tries in a match or for losing by not more than 7. Here is a group table:

Team	P	W	L	F	A	Pts
Apocrifal	3	3	0	83	24	14
Beaugas	3	2	1	28	51	9
Countaphyt	3	1	2	9	16	4
Dewbius	3	0	3	35	64	1

(Here, the F and A columns represent the points scored for and against each team in their three matches: 5 points for a try, 3 for a penalty/drop goal and 2 for a conversion following a try). If there were no drop goals, how many penalties did Dewbius score?

226. Fill in the gaps

(a) ?, M, ?, D, ?, M, ?, D, ?, M

(b) 11, 19, 26, 47, 50, 51, 74, 79, ?, ?

227. Miscellaneous

If CND = 1, NRA = 47, CIA = 86 and CPU = 357, then what is RSI?

228. Identify the following

(a) Song by Blur

(b) Character in *Lolita* and *Ada, or Ardor*

(c) BBC TV series that debuted in 2006

(d) Comic book character, played on screen by Thora Birch

(e) Band who had a no. 9 single in the UK charts in 2004

229. **Which?**

The following list of 55 words can be divided into 10 sets, all of different lengths. Put another way, there is one set of 10 words, one set of 9 words, etc. What is the word in the set of one? See pages 148–9 for explanation.

A	AID	ALEXANDER	ALEXIS	ALIEN
ARM	AURAL	AUTHENTIC	AXIS	BIG
CANDID	CANOPY	CESAR	COL	COMING
DAILY	DEAN	DIFFERENT	EXISTENCE	GALENA
HANSEATIC	INCLINE	INN	INTERPOL	LABEL
LADY	LEANER	LESSEE	LIFE	MARXIST
MILY	MODEST	MOON	NEON	NIKOLAI
OCCUPATION	OMANI	OR	PASSION	PREMIER
PRINCIPLE	RUGBY	SEE	SENSITIVE	SEXISM
SHOOT	SHORE	SINEW	SORE	STRAINED
STUBBORN	SUPER	TAXIS	TIMES	WARN

230. **Missing**

The following set is in alphabetical order, but the ever-unstable kwiz koffee kup has obliterated the penultimate member. All we know is that it had four letters. What was it?

JSYV, KNAJ, MQOPB, POF, VYQ, WKUHH, WRWN, YOD, ****, ZLCLU

231. **Opposites**

The following list consists of pairs of words which are in some sense 'opposites' – and an odd word out. What are the pairs, and which is the odd word out?

BOXING	DOWN	ERGO	GROG	HISS	HOME
MAIL	OGRE	OPPO	RATTLE	REFORM	SILENT
SITES	TILT	TOMTOM	WASH	UP	

232. 5 sets of 5

The words below divide into 5 sets of 5. What are they, and what is missing?

BALL	BET	CLUB	FLU	GENERATION
HINT	ICE	JAY	KELLY	LO
LOBBY	LOCATION	OBSERVE	OX	PEOPLE
PLAN	POT	RATIONALE	SMURF	SYNC
TENACIOUS	TREE	TUN	VAN	YES

233. Crosswords

Across

3 Sign of the zodiac (5)

4 US state (8)

6 Sign of the zodiac (6)

8 Capital city (6)

9 Country in Eurovision 2016 final (6)

Down

1 Number (5)

2 NATO phonetic (4)

4 Element or planet (7)

5 Country (10)

7 Month (5)

What is the sum of 1 down?

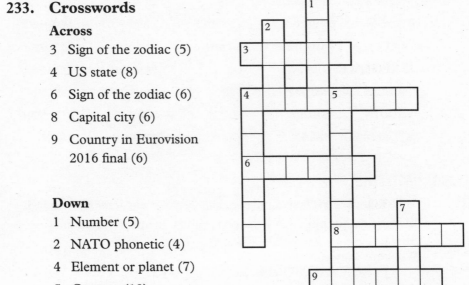

234. Word sequence IX

(a) NOT, LAD, EGG, THE, DEW, HIM, WHO, KEY, JOG, WHY, ?

(b) SO, DIE, MAUL, VOICE, QUIVER, BEEHIVE, SAUCEPAN, HOUSEHOLD, COELACANTH, ?

(c) FOAL, HERD, DRAM, SIDE, TALE, FOND, BLUE, ?

(d) ADDS, ICE, CROP, PECAN, DON, MILLS, LATEX, SULK, ?

(e) ABUT, DANK, CALM, BARK, AGED, SIDE, SHUT, TOTE, VASE, SILT, ?

235. Odd one out

Which is the odd one out: February 2nd, April 4th, June 6th, August 8th, October 10th or December 12th?

236. Cipher

Solve the following (all using the same cipher alphabet, but in various ways):

1. ATGNWC ODC

2. JGL IGQQGACPMHFJ

3. ?

4. VJGN ELUMBAKW

5. VOEMYN ZYNYT PY GKYIION

6. ?

7. ZJKQ HDDHD

8. XUB SV-EKKB

What are 3 and 6?

237. Which name?

If 1=Lash, 2=Captain, 4=League, 5=Neck, 10=Plain and 11=Ted then 6=a proper name beginning with W. Which name?

238. Kwiz kwestion

Various answers to a kwiz kwestion are INDIA (NATO phonetic alphabet), KENT (UK counties), RHO (Greek alphabet) and TEN (names of cardinal numbers). What's the kwestion?

239. Sums

(a) $\dfrac{\text{Myanmar} + \text{Cameroon} + \text{Mali} + \text{Cocos Islands} - \text{Chile}}{\text{Latvia} + \text{US Virgin Islands} - \text{Liechtenstein}} = ?$

(b) $\dfrac{\text{Madagascar} \times ((\text{Australia} + \text{Nicaragua}) - (\text{Sweden} + \text{Cuba}))}{\text{Namibia}} = ?$

(c) $\dfrac{\text{Afghanistan} \times (\text{Burkina Faso} + \text{Afghanistan} - \text{Canada})}{(\text{Germany} - \text{Barbados}) \times (\text{Belgium} - \text{Bosnia \& Herzegovina})} = ?$

(d) If Neon + Iron = Silicon and Cobalt + Ruthenium = Titanium, then:

   i. Bromine + Krypton = ?
   ii. Zinc + Magnesium − Technetium = ?

(e) If Francois Mitterrand = Harold Wilson then who, allegedly, is the following?

   $\dfrac{(\text{Joan Sims} - \text{George Carlin}) \times (\text{Eva Braun} - \text{Eva Peron}) \times \text{Telly Savalas}}{(\text{Peter Boyle} - \text{Nora Ephron}) \times (\text{Ferdinand Marcos} - \text{John Wayne})} = ?$

240. Identify

(a) GGNKBDMMPPODPPOSWLSCNWSSNWD
DSDGDDSBNTNNPINPNGNBSND

(b) GJTJJJAMWJJZMFJAAHRJCSBSWTWTWJHFHDJLRLJRG
WGB

(c) GOCWIKTCEPIPEWNMBCMVFSCFTFLCWPGROCBLC
PSM

(d) VNSFCFTDFGHTDWPPBTFSLFISRMHBFFS

(e) DFGTYODLTTMFOATLGTTDCQSS

241. What follows?

What follows (and completes):

(a) AIRSTOP, BOWLDER, CADENZA, DOUGHTY, EQUATOR, FOREVER, GABIONS, HEXAGON, IMPAINT, JESUITS, KAOLINE, LECTURE, MIDWIFE, ?

(b) HELIUM, BORON, CARBON, SODIUM, PHOSPHORUS, CALCIUM, IRIDIUM, GOLD, ?

(c) 2, 3, 4, 6, 7, 9, 10, 11, 19, 20, 23, 29, 40, 42, 52, 90, ?

242. Name dropping V

The following words used to contain some famous names, before they became stars.

For example: *et cho*ol = Mark Lester (MARKet choLESTERol)

(a) pla*c *isk   (pop star)

(b) fl*l ava*   (author)

(c) *ign la*e   (actor)

(d) *ine to*t   (author)

(e) *id b*er   (actor)

(f) s*et sp*l   (artist)

(g) *ce *o   (actor)

(h) *ty *mp   (actor)

(i) *t s*   (pop star)

(j) *t g*   (pop star)

243. Possibilities

There are multiple possibilities for the numbers omitted (i.e. marked with '?') from each of the following sequences. What are the smallest and largest possibilities?

(a) 1, 51, 550, 502, 600, 900, 153, 1400, 557, 1008, 656, 2007, ?

(b) -, -, 1, 9, -, 2, 4, 7, 11, 5, ?, 20, 99

(c) 7, ?, 5, 1, 2, 82, 100, 97, 54, -, 40, 114

244. Missing scores I

Here are a selection of results from the Krosby International Tournament, sponsored by Koffee Kups UK (slogan: 'you drink, we spill'). Can you supply the missing scores?

SCANDINAVIA			
NORWAY	2	DENMARK	3
FINLAND	?	SWEDEN	3

SOUTHERN EUROPE			
ITALY	1	SPAIN	3
FRANCE	3	GREECE	?

CENTRAL EUROPE			
POLAND	?	CZECH REPUBLIC	7
ROMANIA	4	RUSSIA	3

REST OF THE WORLD			
WALES	4	ISRAEL	?
SOUTH AFRICA	6	ENGLAND	3

245. Chains

(a) Gary Oldman _____ Ethan Hawke

(b) Michael Jackson _____ _____ Patricia Arquette

(c) David Frost _____ _____ _____ Slim Jim Phantom

(d) Debbie Reynolds _____ _____ _____ _____ Laurence Harvey

(e) Lex Barker _____ _____ _____ _____ _____ André Previn

246. A to Z

A = AF, AL, BA, BU

B = GH, JL, PM, RS

C = EC, GB, JB

D = JD, JM, RK, RM

E = AL, DS

F = LH, PM, WJ

G = MR, PC, PG, SH, TB

H = CH, JH, ZH

I = AF, GB, JF, KP, MH, TF

J = BF, PW, RB

K = AF, GS, PC, PS

L = AH, RC, RJ, RL, SC

M = CF, CS, DW, GM, LT, MB, MB

N = DG, KC, KN

O = LG, NG, PA, PM, TM

P = CD, JC, NB, RS, SM

Q = BM, FM, JD, RT

R = BB, MM, MS, PB

S = CD, GL, GT, HK, JB

T = DR, EK, MF, PW, OW

U = AC, DE, LM, PH

V = NM, PS, RA, SJ

W = JE, KM, PT, RD

X = AP, BA, CM, TC

Y = AM, VC

Z = AH, BC, DM, RP, SP

247. Where?

Where does DEN fit in the following list? *(Read left to right, top to bottom.)* See pages 145–8 for explanation.

NEEDLE	BAULK	MAESTRO	GET	DAISY
BADGER	CURSE	FOOD	CROSS	WEDGE
STRONG	SALTY	CHICK	SWORD	SICKLE
TELSTAR	PAIRS	PINK	CHAFF	MORE
MINT	BLUE	PENNY	TICKET	ELITE
NOWHERE	PINCH	BLOODY	MOSCOW	TURNIP
HAWK	CHERRY	SOLO	GLASS	BOFFIN
MAIL	CUCKOO	PEA	HOP	DOG
DEAR	MEDALLION	NET	DOCTOR	WHITE
LOUSE	BRIM	COME	DUET	DAY
UNITS	INSIGNIA	BLACK	BEETLE	BARN

248. Which?

The following list of 55 can be divided into 10 sets, all of different lengths. Put another way, there is one set of 10, one set of 9, etc. What is the word in the set of one? See pages 148–9 for explanation.

AFTER	AGE	BABYLON	BE	BEAR
BLOWN	BLUFF	CHANNEL	COMMON	CONCLUDE
DEW	EGG	EGOTISM	EIGHTIETH	END
FAMOUS	FINISH	GHEE	GHETTO	GIG
GOOD	GROUND	HALT	ITCHY	JUT
KITTEN	LEAGUE	LONG	LORDLY	LOVE
MORNING	NIGHT	PIN	POISON	POST
REAL	SATURN	SHELF	SIX	SLAUGHTERHOUSE
STILL	STOP	TEETH	TERMINATE	THEE
THISTLE	TIGHT	TIPSY	TITHE	TURKEY
VALLEY	VIA	WATER	WET	WILD

249. Name that tune

Explain how the following pieces of music were produced.

(a)

(b)

(c)

(d)

250. Missing scores II

What are the missing scores?

(a)			
AFGHANISTAN	2	GABON	4
PORTUGAL	5	AZERBAIJAN	?

(b)			
BULGARIA	3	KYRGYZSTAN	2
FIJI	3	BOLIVIA	?

251. Elements

Among the elements, what unique property do each of

(a) BERYLLIUM

(b) ALUMINIUM

have? What property is shared by

(c) IRON, LEAD

and no other elements? What property does

(d) TIN

have more than any other element? Finally, what property is exhibited by

(e) LEAD (least, uniquely), PROTACTINIUM (most, uniquely)?

252. Seasonal greeting

The following clue leads to 7 4-letter words – but what is the real (two-word) message?

Weaponry thought to damage, left on boat, dictates that I conquered German river.

253. Number sequence VIII

(a) 441, 961, 691, 522, 652, 982, 423, ?

(b) 2, 5, 8, 10, 13, 17, 18, 20, 25, 26, 29, 32, 34, 37, 40, 41, 45, ?

254. Sums

(a) $$\dfrac{DD + LD + TD}{LaL + SaS + GaL} = ?$$

(b) $$\dfrac{(\text{Mature annuity} \times \text{Inheritance}) + (\text{Bank dividend} \times \text{Repairs on two hotels})}{2^{nd}\ \text{Prize in beauty contest}} = ?$$

(c) $$\dfrac{36°\,26'N\ 120°\,06'W}{36°\,26'N\ 118°\,54'W} \times \dfrac{50°\,59'N\ \ 0°\,39'E}{34°\,48'N\ 82°\,14'W} \times \dfrac{45°\,54'N\ 111°\,33'W}{34°\,44'N\ 118°\,36'W} \times \dfrac{56°\,11'N\ 120°\,32'W}{41°\,48'N\ \ 86°\,37'W} \times 34°\,48'S\ 173°\,00'E = ?$$

255. Quelle lettre manque?

Pourquoi?

(a) V, I, V, I, M, V, I, L, M, I, I, I, L, L, M, C, V, M, ?, C, M

(b) V, D, M, I, I, M, I, I, M, I, I, V, L, V, I, V, M, I, L, L, L, I, I, L, M, I, D, M, I, D, ?, I, D

256. Explain the order

2018, 2024, 2087, 2076, 2016, 2021, 2019, 2023, 2020, 2017, 2025, 2022

257. Identify (precisely)

(a) FJTEPA, GSEO, PJAAE, RBNA, RJFWE, YABBFW

(b) BDCE IPID, ECWWDK, KDTHBTDK, KHID, MICGGDK, UCGEBDMQPUS

(c) CICCIA, GLJJY, HBJBGIJ, KTG, NLWBNTK, OJBFB, TJTWBJA, WIGNI, WTAFTNO

(d) IHN KTHEMNE DHE BIMQX BITM, QKRMQNNH, QVNGUN, QVNHQX

258. Scrambled lists

Each of the following is a complete set. The number of items in each set is given.

(a) CCDEEEEEEFGHLOOORRRRRSSTTUW   (3)

(b) AAEEHHJKKLMMNORTTUW   (4)

(c) AACDDEEEEGHHHHIIMNNNNOORRSSSTTVWWYY   (5)

(d) AAAACCDEEEEEGHIIIIJLLLLLMMNNNOPPSS   (6)

(e) AAACCCCDDEEEEEEEEEFFIIIIJHHHMNNOOPPPRRRRST TTTTTUUUY   (7)

259. Word sequence X

What word could follow:

(a) AIR, ART, ABILITY, AVER, ANGLE, AMBER, ACRE, AT, ACTOR, ?

(b) DRY, LIE, HIT, RYE, ROB, BAR, GET, YEN, NIL, ONE, ?

(c) LAP, ROB, AIL, LAD, HOE, TOO, FOG, LET, DIN, LIE, ?

(d) AFT, BEL, COX, DRY, EMU, FIR, GNU, HIM, IMP, JOY, ?

260. Next?

What (cipher and plain text) will follow:

KGEL, QGBWG, FYVDOH ODRW, FSGDLA, FHGRNYXE, FHQOHU, EHQ XGPYEYQ, NLGRC, VXKOYCGFX, XQCXFQX, NWOFLW, XQPLFN, IYDBDGP, EHGYHG, ?

261. Kinema

Inspired by a trip to the Kinema, the Setters set some Krossword Klues:

(a) B (3, 5, 7)
(b) £6000 (6, 7)

Can you identify the films?

262. Next in sequence

Identify the items marked with a '?'. Note that '…' represents further entries that you are not expected to provide. The values are exact.

.349, 9.3, 529, 5.433, .974, ?, ?, …

263. Miscellaneous

(a) Explain the following:

    i.   TB/SBM, SK

    ii.   SA/SL, TK

    iii.   NL/TGC, PP

    iv.   RD/M, TH

    v.   CFTD/TDA, JLC

    vi.   DADOES/BR, PKD

(b) If AB=dispute, HP=trickery and NP=insipid, then what are:

    i.   FD

    ii.   WN

    iii.   pudding

    iv.   musical instrument

(c) Pair up these first names:

Barbara, Basil, Dale, Doug, Jane, Nick, Oscar, Patti, Roxy, Tamzin

with these surnames:

Adams, Bond, Cooper, Donnelly, Ellis, Gentry, Norman, Oldfield, O'Neill, Vernon

264. Next term

In the following sequence, what might be the – perhaps inappropriate – 5-letter 11th term?

AMUSING, EXODUS, HISTORY, EQUATOR, QUICKEN, SPHINX, OPPOSITE, AZIMUTH, EFFLUENT, INDEX, ?

265. A hard question?

Here's a harder kwestion. Identify and place the following in the correct order:

BWXIH, FDOFLWH, FUFYNYJ, IARUWO, JPYSVMXI, LXADWMDV, NSKWYXN, UBMD, UXZNUIRGYK LKRJYVGX, XBHYAG

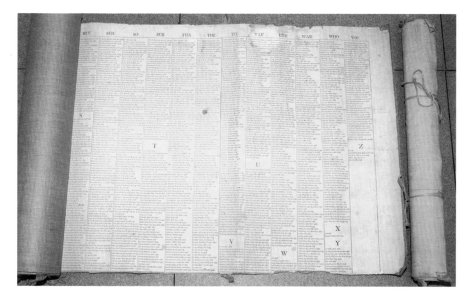

Codebooks are a way of managing lists of words and phrases, and the codes that represent them. They are normally printed on paper and bound as books. This unusual 'Coderoll' from 1820 has been transcribed on to cloth so that it could be hung on a wall and used simultaneously by several clerks if necessary.

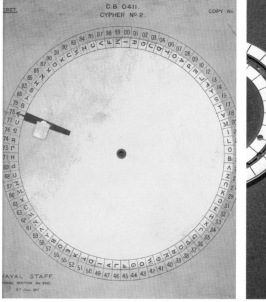

A codewheel used by Royal Navy ships to encipher and decipher tactical messages in 1917.

A WWI cipher machine designed by James St Vincent Pletts.

The charter issued in 1914 by Winston Churchill (then the First Sea Lord) for 'Room 40', an early predecessor to GCHQ, and the cryptanalytic unit of the British Admiralty in WWI.

This is the earliest photograph of members of a British Signals Intelligence (Sigint) Unit, and shows staff who worked at Leiston, Suffolk. Among them was William Swanborough who joined the Signals Service of the Royal Engineers in 1918.

Sergeant Swanborough, pictured front left, stationed in Kabul in 1920, where he was tasked with intercepting Soviet signals. Swanborough went on to become Wing Commander and Commanding Officer of the sigint station at RAF Cheadle during WWII. He was in charge of the station as an officer and a civilian for a record 37 years.

A German message intercepted by a British operator stationed at Basra in 1918. The note at the bottom says that 2 letters were missed due to the machine jamming.

The elegant Kryha Machine, used for encryption and decryption from the 1920s to the 1950s.

This Enigma, A320, was the first obtained by GC&CS (the Government Code and Cypher School, GCHQ's name from 1919-1946). It was purchased in Berlin by the Deputy Head for £30 in 1926.

Hagelin was a Swedish company that produced cipher machines. The B-211 (detail above) is the oldest Hagelin machine in GCHQ's collection.

II. Cambridge: [Continued]

E.J. Passant	M.A.	Sidney Sussex College, Cambridge	
J. Saltmarsh	M.A.	King's College	"
F.H. Sandbach	M.A.	Trinity College	"
P.F.D. Tennant	M.A.	Queen's College	"
A.M. Turing		King's College	"
W.G. Welchman	M.A.	Sidney Sussex College	"
L.P. Wilkinson	M.A.	King's College	"
M.H.A. Newman		St. John's	"
C.T. SELTMAN		Queens	

List 2.

W.L. Cuttle (Modern Greek)	M.A.	Downing College	"
C.W. Guillebaud	M.A.	St. John's College	"
C.J. Hamson (Turkish)	M.A., LL.M.	Trinity College	"
R.J.H. Jenkins	M.A.	Emmanuel College	"

In 1937, GC&CS began to think about how to recruit a new sort of cryptanalyst and, in collaboration with WWI veterans, drew up lists which contain some of the greats who would work at Bletchley Park during WWII.

	March 27th	March 28th	March 29th	March 30th
D.J. Allan (2)	Commercial	Commercial		
Dr. Beeston (3)	Near East	Near East	Near East	Near East
(2) Professor A.H. Campbell	Air	Air	Military	Military
(1) J.M. Dawkins	Near East	Near East	Near East	Near East
(2) L. Forster	Naval	Naval	Naval	Naval
(3) Professor Fraser	Mr. Turner	Mr. Turner	Mr. Turner	Mr. Turner
(1) E. Lobel	Research	Research	Research	Research
(2) Professor Norman	Military	Military	Air	Air
(2) Dr. E.G.C. Poole	Research	Research	Research	Research
(3) F.A. Taylor	French	French		
(2) Professor Tolkien	Scandinavian	Scandinavian	Spanish	
Professor Waterhouse	Naval	Naval	Naval	Naval
(2) Professor Willoughby	Naval	Naval	Naval	Naval

J. R. R. Tolkien was identified as a potential cryptanalyst in 1939 and took part in a couple of training weeks. In March 1939 it was noted that he was keen on the work but in the event was not selected for Bletchley Park, where the need at the outbreak of war was for mathematicians rather than linguists.

RAF Cheadle, a British sigint 'Y-station', during WWII. Staff at this site, mainly women, intercepted enemy communications.

349 hrs   ror 1106/6/++

      roo 1130

pp 202

From: ·R 221

The BÉNOUVILLE bridge over the CAEN canal is in the hands of the British. Two gliders have landed. Am withdrawing after a brush with the enemy.

1210/6/6/++

'Headlines' were a quick way of getting short pieces of intelligence to military commands. This one, timed 14:10 on the 6th June 1944, records a German officer informing his headquarters of the capture of the Bénouville (Pegasus) Bridge on the morning of D-Day.

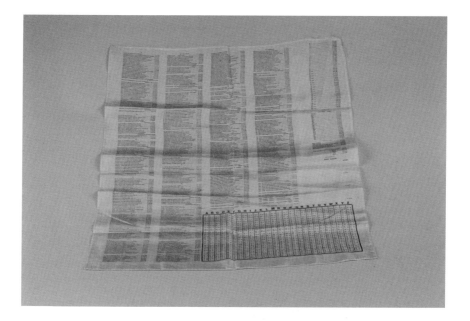

Sensitive material was often printed on to silk, so that it could be easily scrunched up and concealed by agents sent into occupied Europe.

The famous Bletchley Park was run by GC&CS during WWII as a site for codebreaking.

'Morrison Walls' were a way to map enemy radio networks. They showed the location of transmitters, the way they communicated with each other, and the type of traffic they sent.

Left: Alan Turing was a mathematician employed by Bletchley Park at the outbreak of WWII. He pioneered the use of 'bombe' machines to decipher German messages encrypted on Enigma machines (below) on a mass-scale. This breakthrough saved countless allied lives.

Expanding $(1-x)^{-3}$ by the binomial theorem

$$(1-x)^{-3} = 1 + 3x + 6x^2 + 10x^3 + 15x^4 + 21x^5 + 28x^6 + 36x^7 +$$
$$+ 45x^8 + 55x^9 + 66x^{10} + 78x^{11} + 91x^{12} + 105x^{13} + 120x^{14} + 136x^{15}$$
$$+ 153x^{16} + 171x^{17} + 190x^{18} + 210x^9 + 231x^{20} + 253x^{21} + 276x^{22} + 300x^{13}$$
$$+ 325x^{24} + 351x^{25} + 378x^{26} + 406^{27} + 435x^{19} + \ldots$$

Now multiply by $1 - 3x^{10} + 3x^{20} - x^{30}$ and we get

$$f(x) = 1 + 3x + 6x^2 + 10x^3 + 15x^4 + 21x^5 + 28x^6 + 36x^7 + 45x^8 + 55x^9 +$$
$$+ 63x^{10} + 69x^{11} + 73x^{12} + 75x^{13} + 75x^{14} + 73x^{15} + 69x^{16} + 63x^{17} +$$
$$+ 55x^{18} + 45x^{19} + 36x^{20} + 28x^{21} + 21x^{22} + 15x^{13} + 10x^{24} + 6x^{25} + 3x^{26} + x^{27}$$

Part of a calculation by Alan Turing.

Above: The factory in Letchworth where bombes were manufactured for code-breaking. These bombes were located at outstations such as Stanmore and Eastcote and given codenames based on places in the British Empire and beyond. They were then controlled from this room (below) at Bletchley Park.

Working aids were notes or equipment used to speed up complex tasks, particularly for working on Enigma decryption. They were often makeshift, such as the one above, created from a typewriter platen.

Card indexing was an important part of the analytical process at Bletchley Park. This card references intercepted messages to Wilhelm Canaris, the Head of the Abwehr (German Intelligence), and cross-refers to his codename Guillermo.

Bletchley Park worked at the cutting-edge of technological design, and a large number of pieces of equipment were designed and built from scratch to support cryptanalysis during the war.

sti-ff, en, ness, of, the—9619
stif-l e, the, stigma, tis e, of, the—8148
stil-l, be—4855
still further, to—9187
—— less, to, stimulant, to, the—9200
—— more, to, stimulat e, the—7607
stim-ulus, to, the, sting, stingy—6033
stin-t, of, the, stipend, iary, of, the—9280
stip-ulat e, ation, that, the—6603
stir, up, the, Stock Exchange—5493
sto-ck, of, the, Stokes, 's, Mr.—6405
stoi-c, al, stoker, s, stole, the—9628
stol-en, the, stolid, stomach—4293
ston-e, y, stood, stoop, to—5645
stop, stoppage, of, the, Store Department, 's—8796
stor-e, storm, y, of, the—9305
store-s, stor y, of, the—9245
stou-t, ly, stow, Strachey, 's, Sir R.—8790
straight, to, the, straggl e, er—8327
—— line, to, the, Straits Settlements—9207
str-ain, ed, the, Straits, of—6215
stra-it, en, ed, strand—8763
stran-ge, stranger, to, the, strangl e—7406
strat-agem, of, strategic, Stratton, 's, Maj.—7584
strate-g y, ical, stratum, of, the—7505
straw, stray, stream, street, of, the—4645
stre-ngth, en, of, the, strenuous—8458
stres-s, on, the, stretch, strew—4572
stri-ct, ness, of, the—3756
stric-tly, stricture, stride, of, the—3484
strif-e, strik e, ing, of, the—3850
strin-g, of, stringent, orders, to, the—8948
strip, stripped, of, stripe—7993
striv-e, to, stroke, of—3875
stro-ng, stronghold, of, the—7900
—— est, strove, to—6134
—— ly, struck—6251
stru-ggl e, for, the, strung, structur e, of—8146
Stu-art, 's, Mr., stubborn, stuck—7366
stud, stud y, student, of, the—6632
studi-ous, ly, stuff, of—4781
stul-tif y, ication, of, the—4731
stum-bl e, stumbling block, of, the—6747
stun, stung, stupendous—8091
stup-ef y, ied, action, at, the—8995
stupi-d, ity, of, the—6878
stur-d y, ily, iness, of, the—6649
sty-le, of, the, Suakin, suave, Subadar—8964
sub, sub judice, subaltern—8759
su-bdivid e, the, subdivision, of, the—9983

179

subd-u e, ed, the, Subiya—0884
subj-ect, of, the—9625
—— to, the, subjection, of, the—0942
—— to your, subjoin, ed—1064
—— s, of, the, subjugat e, ion, of, the—1930
subl-ime, submarine, submer ge, sion—2194
subm-it, to, the, submissi on, ve—1369
subo-rdinat e, ion, of, the, suborn—1022
subp-œna, subscrib e, er, to, the—2641
subs-cription, to, the, subservient, to, the—7126
subse-quent, subserviency, to, the—2275
—— ly, subsidence, of, the—2334
subsi-d e, subsidiary, subsidies, to, the—0944
subsid-is e, the, subsidy, of—7182
subsis-t, ence, of, the—0801
subst-ance, subsistence allowance, of, the—2308
substa-nti al, ally, substantiat e—1242
substan-tive, appointment, of, substratum, of—1149
substi-tut e, ion, of, the, subterfuge—1331
subt-le, ty, of, the, subterrane an—3063
subtr-act, ion, suburb, an, of, the—4056
subv-ention, subver t, sion, of, the—1094
Subyeh. See Subiya.
suc-ceed, in, subversive, of—1070
—— to, the, succeeded, in—1775
succe-ss, of, the, successfully—1659
—— ful, in, succession, to, the—2305
succe-ssive, ly, successor, to, the—1132
succi-nct, succour, of, the, succumb—0948
such—1184
—— a, n—1097
—— are, the—1286
—— as, to—2870
—— be, the—0802
—— is, the—3124
—— manner as, to, suck—8044
sud-den, ness, of, the, Sudder—4641
sudd-enly, Suez—3176
sue, the, Suez Canal—2948
suf-fer, er, from, the, sufferance—7143
—— ing, suffic e, iency, of, the—3618
suff-icient, to, suffocat e, Suffolk, Regiment—1047
—— ly, to, suffrag e, an—1396
sug-gest, that, the, sugar—0896
sugg-estion, of, the, Sugar Bounties—1102
—— ions, of, the, suggested, that, the—1199
sui-cid e, al, suitability, of, the, suitor—1197
suit, suitab le, suite—2834
Suk-kur, sulk, iness, y, of, the, suitor—1847
Sul-iman, sullen, ness, sull y—0981

180

A page from a codebook used to encrypt messages between King George VI and Pakistani Governor General Muhammad Ali Jinnah in 1947-48. Each number sequence has several words that it could relate to, which would become clear in the context once decoded.

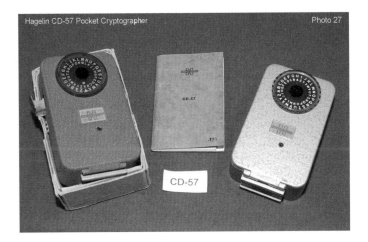

The Hagelin CD-57 pocket machine was a very portable encryption device designed in 1957 so that users could encrypt messages when on the move.

One of two 'NOREEN' cipher machines used on board Her Majesty's Yacht Britannia from the 1960s.

The modern day face of British sigint is GCHQ, based at 'The Dough-nut' building in Cheltenham.

GCHQ's mission is to keep Britain safe and secure, working with MI5, MI6 and law enforcement. Its headquarters is based in Cheltenham, with regional hubs in Scarborough, Bude (pictured), Harrogate and Manchester.

Inside GCHQ's main building in Cheltenham.

266. Elementary crossword

Across

2 Intersection of 16 across and 17 down (7)

6 Device for raising the pitch of a guitar (4)

7 Axed cabinet minister, we hear (4)

8 England footballer or chief banker (6)

9 Dee, or Katharine, or Audrey (7)

10 Invented by Da Vinci, allegedly (10)

14 Employer of 15 down (3)

16 Britpop band (5)

19 Way of speaking (8)

21 Supernatural dramatic device (3 or 5)

Down

1 Alpha gorilla (10)

2 Obscure (6)

3 Indian soldier (5)

4 'Save the __, save the world' (11)

5 Hellfire (7)

9 20p or 50p (8)

11 Tennis player (2,2)

12 What happens in Vegas (and Miami, and New York) (3)

13 Greek letter – sounds easy (2)

15 Employee of 14 across (3)

17 Rise (6)

18 Prava, or Simva, or Atorva (6)

20 Landlocked country (4)

267. Number sequence IX

(a) 16, 17, 20, 22, 24, 31, 100, 121, ?

(b) 1, 4, 10, 17, 27, 40, 54, 71, 100, ?

(c) 1, 4, 9, 10, 19, 24, 31, 40, 51, 64, 79, 90, ?

(d) 1906, 772, 304, 130, 82, 52, 34, 22, 16, ?

(e) 1010001, 100111, 21301, 20141, 15041, ..., (14642nd element)

268. Sums

(a) $\sqrt{(\text{Cleek}^{\text{Cleek}} + \text{Mashie}^{\text{Cleek}} + \text{Mashie-Niblick})} = ?$

(b) -----.---.. + ... = ?

269. Which words?

Each of the following sets leads to a word. In each set one member suggests the first letter of the word, two members suggest the second letter, three members suggest the third letter (etc. for the bigger sets). What are the words in each case?

(a) CLAUDIUS, DRONE, DUBZ, HONEY, ROBOT, STING

(b) GARDEN, PAD, POD, SPLIT, SWEET, ZULU

(c) AL, APSE, ATE, BOW, CHEPSTOW, EVEN, EVIAN, FIN, GAR, HAY, HEREFORD, IDE, MONMOUTH, PERRIER, ROSS

(d) BAG, BLACK, BRIGHT, CHICK, COSY, CUP, DEAR, EVIL, LEAF, LENGTH, NO, POT, PRIVATE, SHORT, STIFF

(e) BAY, BOAT, COLI, CRIME, GOVERNMENT, HORSE, LION, MAIL, OBSERVE, OGLE, SHORE, TURN, VIEW, WALKING, WEED

(f) ATTEND, AUTHOR, BELL, COUNT, DUB, EWE, GIANT, JOY, LESS, PRIEST

(g) A, BONEY, DOVE, EMCEE, FISH, FOR, IN, LOTS, MARKS, MINUS, MORE, OF, OF, ONE, SPOT, SQUARED, SQUAREROOT, THE, THE, THE, WINGS

(h) AGE, AGEISM, BET, BIT, DEATH, DECAY, ENGLAND,
EVACUATE, GUATEMALA, GUINEA, IMAGES, LABOUR,
LESION, MAGICS, MALE, NUMERIC, PARTICLE, PONIES,
RADIATION, RHYTHM, SCINTILLA, SMIDGEON, SPECK,
SQUAT, STIGMA, WHIT, YEAR, YORK

270. Oetfn ciameld

It is oetfn ciameld taht a snteecne or prshae is utdsaednlnrbae eevn if
olny the fsrit and lsat ltetres of ecah wrod are in the rghit pcale and the
rset jbumeld. u ah ubo eh h rsi n sa retet r sinsi n h se lit mbuel? ah re
h wollnio ingillar?

(a) n w reh uo vi

(b) deiaxlirisceupupitiolcriciafgalo

(c) o va u cioaur eue

(d) tn ca fi mo ia su la

(e) l o n n n o l

(f) o

271. Answer the question

OPEN, FEATHERS, TABLES, LEGS, SIGN OF THE, CARD
TRICK, FOURS, QUESTIONS, CLOSE, A COMMON
LANGUAGE, IT TAKES, FIRST AMONG, ?

272. Identify the following

(a) Lyman FB	(e) George GL
(b) Manoj NS	(f) Fahrid MA
(c) Joseph AR	(g) Francis SF
(d) Lafayette RH	(h) Robot DO

273. Which number

Which well-known number is described here?

```
WII SFCKU TI YPN JRUEXUJKTKRFH WI C JRWCNM BS JCZ
EOJPNCHY
```

274. Where?

Where does ENTRY fit in the following list? *(Read left to right, top to bottom.)* See pages 145–8 for explanation.

WOLFIE	GORE	MOODY	DRAM	FILCH
LAST	PAIR	CROUCH	PREPARED	BOYFRIEND
ARCHIE	OFFICE	TEA	CASTLE	ADO
CUT	FUDGE	AWARE	DIE	GOLDEN
VIKING	PINA	FOE	CAKE	EVER
COUNT	MAN	TAILS	LOVE	FLAT
LIPSTICK	MATADOR	BROWN	BREAK	LAND
LESS	HIRE	NEW	PLAIN	BAG
SHOT	LUPIN	SWIM	BLACK	LATER
SCENE	POWER	SEE	RIDDLE	TREAT
BARNEY	TON	SPIDER	SIN	GOOD

275. Sort out the characters

Identify the ten people in the following list:

AAADDEEEEEEEEFFGGHHII
JKLLMMMNNNNNNNOOO
OOOORRRSSSSSTTTVWW

276. **Which?**

The following list of 55 can be divided into 10 sets, all of different lengths. Put another way, there is one set of 10, one set of 9, etc. What is the word in the set of one? See pages 148–9 for explanation.

AGE	BARD	BART	BETTER	CALCULATE
CHIEF	COMMAND	COURT	DIAGRAM	DIGIT
DIVE	DIVERT	DOG	DRUID	ELBOW
ELDER	EMERGE	ERODES	ERR	ETON
FISHMONGER	FORM	GALLOW	GAUL	GLOVES
GNAT	ILLUSTRATION	JACKET	LAKE	MERCHANT
MINCE	NESSIE	NIECE	NUMBER	OGIVES
POORER	RECKON	REEF	REGIME	RICHER
RUGBY	SEA	SESTET	SHAPE	SMITH
SONG	THEATRE	TROUSERS	UPPING	WAISTCOAT
WILLIAMS	WINCHESTER	WIT	WOO	WORSE

277. **Divide into sets**

ADO	AGENT	COAST	CREST	CUP	FINCH	FINGER
FULLY	GARDEN	GEORGE	GUIDE	LINING	PLATE	POLICE
POWER	RIDE	RUSH	SANTA	SCOUT	SEVEN	SQUIRREL
SOCIETY	STICK	STONE	SURFER	TALK	WONDER	ZONE

(a) Divide the words above into sets

(b) Which bird?

278. Identify the following

Note that '...' is used to indicate that we have provided only the start of a longer sequence of words.

(a) WHEN ARAGORN PIPPIN

(b) MR NOT HARRY THE THE IT THE

(c) THIS ONCE THIS ONCE THERE IT IN

(d) THE THERE THE WITH THE PUNCTUALLY JAMES ...

(e) I HE'S MOMMY

(f) THE LANDSCAPE-TONES AS THE

(g) IN THE THIS THIS THERE THE NOTHING THIS THE WATCH THE THIS ...

(h) IMAGINE THE IN HITLER

(i) WHETHER DR IT THE IT NO

(j) THE THE ONCE WE CROSSING ALBERT SNOW WHEN TOWARDS REVERTING THE DUCK

279. Word sequence XI

(a) EXACT, DHOBI, ANGLE, RANCH, GIVEN, INGLE, THEIR, NEIGH, EXTRA, ?

(b) BLAZER, WILLOW, LOCULI, HAMPER, ASKING, INDIGO, CRUNCH, TARTAN, KNIGHT, ?

(c) WANDER, BREATH, LYRIC, NAIAD, GRADE, FARED, REGARD, THREES, ?

(d) REMERCY, SEVEN, TRACE, RAMP, PICTURE, HAUNTS, UNSURE, ?

(e) MAGNUM, SABBATICAL, ICELANDIC, HEADBOARD, MALEVOLENT, FRATERNAL, GASSED, PHANTOM, DELICATE, ?

280. Sequences

Please identify the elements marked by ?. Sequences (c) and (d) are endless.

(a) T, T, S, S, RM, PM, ?, ?

(b) A, A, A, A, A, E, J, A, ?, ?, ?, ?

(c) E, O, E, F, E, I, E, ?, ?, ?, ...

(d) O, E, E, E, ..., E, E, E, A, A, A, A, ...

Sequence (d) starts with a single O, then has a run of E's of moderate length (between 10 and 100), after which the sequence consists entirely of A's. Exactly how long is the run of E's?

281. Next triangle

(1,2,3), (8,9,10), (10,11,12), (11,12,13), (?,?,?),

(a) What is the next triangle (consecutive integers only)?

(b) In like manner, what is the first quadrilateral (consecutive positive integers only)?

282. The Coronet

On yet another hunt through the kwiz archive for possible ideas for kwestions the Setters came across a 7×7 grid. The words THE and CORONET had been picked out in the grid.

		E				
		T	H			
			O	R	O	N
	C					
		T			E	

(a) What is the missing 5-letter word and where does it appear in the grid?

(b) Why is the grid an appropriate size given its contents?

283. Simple substitution

The kwiz setters enjoy using simplw suhstitution eokws rn quwstrons. Rn tsrs wxtypow, stk rt hwwu uwewsstry, wsres owttwl wguok stvw hwwu dbwk igl Z?

284. A to Z

A = VDT		N = JDI
B = TTOP		O = TFBTFO
C = AWGAXFAE		P = PPPPUAP
D = KNNPOAKG		Q = IALLBTAH
E = IKGAGAG		R = DYLAETGTYLR
F = EWDIDBY		S = SYKW
G = TBAMCG		T = YOGAOWT
H = RTPOBCR		U = MATSOB
I = SFASP		V = IOEILWARAAV
J = DDIDI		W = EOOTCPI
K = HABHAK		X = TDC
L = BIWI		Y = LYFDTW
M = AMADHYWRAP		Z = TAOS

285. Two 3s, two 5s

There is a well-known puzzle in which you are required to construct as many integers as possible using just four 4s and the operators $+ - \times /$, together with brackets, decimal points, factorials, roots, exponents and recurring decimals. Thus:

$12 = ((4! - 4) \times .4) + 4$

$32 = \text{sqrt}(.4) \times (44 + 4)$

And $63 = (4^4 - 4) / 4$

In this version you have all the same rules, but the only digits used must be 3, 3, 5 and 5. So, for example:

$10 = (35 - 5) / 3 \text{ or } 35 / 3.5$

Construct all the numbers from 180 to 189 inclusive. It is acceptable not to use all four digits, should it be possible.

286. Some sums

Although at first glance the equation $2+3+6+9 = 1+4+7+8$ doesn't seem particularly exciting, it is slightly more interesting than your average sum. $2+3+10+11 = 1+5+8+12$ is a little more interesting, and $2+3+9+13+16+20 = 1+5+8+12+18+19$ is yet a little bit more interesting.

Find a similar equation that is even more interesting. (You shouldn't need to use any number bigger than 35).

287. Cherchez le mot

```
A  V  E  R  C  A  N  A  R  D
N  A  N  I  S  W  E  C  U  P
G  R  E  N  O  U  I  L  L  E
U  E  R  S  I  S  H  F  R  R
I  N  O  H  C  O  C  O  C  R
L  E  E  N  C  A  H  R  E  O
L  E  G  N  I  S  R  I  R  Q
E  W  O  O  A  R  E  G  F  U
P  I  N  G  O  U  I  N  O  E
L  D  C  O  R  B  E  A  U  T
R  A  T  G  R  I  L  L  O  N
```

288. The Tree of Partial Knowledge

```
            F
           P V
          A A P
         A E F W
        A D G J T
       A B D E E G
      B D D G H S S
     C F G I J R U U
     C C E E M P T T U
    E F F N O S S T T T
   B B C C H H M P S W
  A A D F J J J M M N O S
 C D G M M N N N N P R S V
            | |
          _| |_
           \   /
            \_/
```

289. Sums

$$\frac{(\text{Cadmium} + \text{Vanadium} - \text{Chlorine})}{\text{Lithium}} - \frac{\text{Cadmium}}{\text{Carbon}} = ?$$

290. Word sequence XII

(a) STRONG, HEARTY, POPULAR, INTELLIGENT, OVERAMBITIOUS, ?

(b) SPRY, STRONG, HEARTY, POPULAR, INTELLIGENT, AMBITIOUS, ?

(c) LOT, CHE, CAT, SHE, LAR, SAY, ESS, LEE, ASP, RED, CIT, REE, ?

(d) SQUELCH, PARTING, TUMBLED, MARXIST, EMPTILY, SPLAYED, DETAILS, INSTEAD, ?

291. Quelle est la prochaine?

MLC, RVM, QCNSS, EKSN, BFVA, QFW, QAVAK, AECDR, LGLA, SBM, ?

292. Wordsearch

Here is a wordsearch, beginning (4, 7, 6, 4, 9, 7). Actually it's a letter-search – but which letter?

```
G  D  S  O  E  A  N  S  N
Q  N  T  K  H  M  B  E  U
R  I  I  I  A  R  E  D  A
E  E  L  N  M  O  F  F  M
H  A  R  L  I  S  O  O  N
T  F  S  Q  U  A  R  E  O
O  O  E  G  N  F  M  E  O
N  R  S  R  E  T  T  E  L
A  R  A  T  G  K  W  E  R
```

293. Miscellaneous

(a) If Taurus = Mail, Sagittarius = Deer and Pisces = Mary, then Cancer = ?

(b) If $A - A = M$, $S - S = J$, $C - U = P$, $HB - M = N$, $EB - U = D$, $E - WR = C$, $B - WC = V$, then:

   i. $B - H = ?$

   ii. $B - W = ?$

   iii. $(E+C) - (H+W) = ?$

294. Easy question

Now an easy question. What property do all the following words have in common?

BACON, BOWLS, BURNT, CANDID, CORPORAL, GALLON, LAND, POT, QUALITY, QUIT, RATIONAL, RED, SILICON, SLIGHT, SPY, SWINGING, TRADING, TWIN

295. Kwiz krossword klues

Here are 24 crossword clues. But what is the answer to the missing 25th clue?

a. Distort	e. Spouts	i. Restaurant	m. Gown	q. Not that	u. Surfeit
b. Water	f. Test	j. Greedy	n. Greasy	r. Rent	v. Floor covering
c. Wreck	g. Mexican snack	k. Fibonacci number	o. Rainbow colour	s. Persia	w. Cancel
d. Painful emotion	h. Smoky fog	l. Paradise	p. One who observes narrowly	t. Dispatch	x. Sound the horn

296. Missing person

Identify the missing person (description in parentheses). The lists aren't complete.

(a) T, NG, TTT, TNTT, COONU, TFTNGT, TOUNDCD, TYVOFNDO, TCNTOFVNC, NTONYNDCOT (author, full name)

(b) DO, OD, OOK, OOY, VDD, HUDB, DODOV, OHJYV, OUWHOV, HWHHODU (author, full name)

(c) D, SY, CCK, GLY, NAK, THY, QMFS, TMWV, UDBLL, FYUYSNLY (unclued actor, full name)

297. Where?

Where does 507 fit in the following list? *(Read left to right, top to bottom.)* See pages 145–8 for explanation.

213	659	855	320	607	673	359	850	600	588	090
525	483	731	366	361	551	328	795	423	425	181
650	400	360	045	160	315	800	297	120	500	280
504	700	626	100	200	962	432	180	135	225	752
886	300	232	240	400	998	200	270	060	263	420

298. Identify

Identify, giving the appropriate phrase *('-' is used for a string containing no characters).*

(a) BLZ, C, COM, CUP, H, O, PC, VX

(b) -, A, AA, ADA, AM, BM, D, M, MY

299. Final word

What is the final (sixth) word in these sequences?

(a) CLOUTS, OBERON, DANGER, PEOPLE, BEMOCK

(b) BARBED, SCHUSS, AGENCY, HOVERS, TERROR

(c) PREFAB, ARTFUL, KEEPER, INCITE, MANIOC

(d) MILLED, ONWARD, STANDS, WAIVES, ALINES

Too straightforward? Try these.

(e) CLUMSY, BUCKET, BROKEN, JARGON, DIADEM

(f) ATRIUM, RENTAL, GLANCE, TRIPOD, SUNDAE

(g) KOSHER, OOMIAK, ENDING, ORPHAN, MISHIT

300. Pentanomes

What's the next 5-digit group?

(a) 11213 12415 12361 71248 13912
51011 11234 61211 31271 41351
51248 16117 12369 18119 ?????

(b) 11111 11555 15115 11111 01010
11011 10111 10151 05105 11051
11051 11101 10101 01010 ?????

(c) 20213 94356 62728 78910 61091
28135 15115 91701 84193 ?????

301. Fractions

Here's a complete set of fractions in ascending order (with '*' denoting
fractions that have been reduced to their simplest form, e.g. 400/250
would become 8/5, 400/200 would become 2). What's the missing
fraction?

1/17(*), 3/50(*), ? , 23/220 (*), 8/75(*), 17/150 , 3/25(*), 23/5,
53(*)

302. What's the common property?

CARDINALS:	THIRTY ONE
COUNTRIES:	MOZAMBIQUE
ELEMENTS:	SILVER
NATO ALPHABET:	CHARLIE
US STATES:	MAINE

303. One more

Below are enciphered 16 members of a set of 17. The first letters of all 17 (unenciphered) can be arranged to form 'EGG MAKERS FLY? MARCH!' What is the enciphered form of the missing 17th member?

DXTTRF-XDXS, IRDMT, IRYOVRKKXI, LMWXTTMAHY, LMYMIRAH, MCIHYMA, MSXTHX, NAMIXN, OHAW, THGGTX, VULZRTSG, WMTMKMWRN, WXAGRR, XIXYG-YIXNGXS, XLKXIRI, YVHANGIMK

Kristmas Kwiz Challenge

Welcome to the Kristmas Kwiz Challenge. This is a chance for you to test yourself against the puzzle solvers of GCHQ by attempting the kwiz under the same conditions as they do.

Each year, for over twenty years, the Kristmas Kwiz is issued in the second week of December. Answer sheets have to be submitted by a date towards the end of January, which means entrants have about seven weeks to solve the puzzles. A few people enter individually, but most entries come from teams – which must be limited to at most four people.

What follows is essentially the 2014 Kwiz. Your challenge is to approach it as it is meant to be approached. Allow yourself seven weeks to solve the puzzles. Or get together with friends to form a team, and see how you get on over seven weeks.

In the kwiz each kwestion has an associated score, which should help with indicating difficulty. The total score is 100.

You will not find the answers to these kwestions in the back of this book. The answers will be available from www.gchqpuzzlebook.co.uk from 1 April 2017.

To avoid spoiling other people's fun, please do not post answers to these questions on the internet, or use public discussion forums to discuss the questions and answers.

1.   Let's start with a bang!

> *Six hours hath September,*
> *April, June and November,*
> *All the rest have four more,*
> *Excepting February alone,*
> *And that has – how many minutes?*

2. Where does ROCK fit in the following list? *(Read left to right, top to bottom.)*

CAMP	KING	CURRENT	CHAIR	STINKING
FASHION	HOLLY	LUMINOSITY	KNOWLEDGE	BACK
PLEASURE	BALLET	GRACE	SCARE	SAGE
DRAIN	EDDY	CAKE	VIE	Y
LANCE	DOUBLE	LUNGED	COMBAT	SCIENCE
BREAM	TEMPERATURE	MASS	MUD	RED
RAIL	REED	INMATE	LENGTH	AMOUNT
COMMERCE	YOUNG	POSE	GREEN	POLITICS
GUARD	CROSS	WORSHIP	STEAK	TIME
STEEL	WHITE	CURSE	LAC	PAT
FORCE	BLUE	NOW	WEALTH	HAM

(5)

3. *This style of question commemorates* Round Britain Quiz *on BBC Radio 4: the format is one, long, cryptic question which has six parts to it, indicated by the letters a–f. To gain full marks you should identify all six parts. This will be sufficient to answer the question.*

An officer of the law who retaliates badly (a), an unseeing feline mariner (b), a seemingly haphazard deliverer (c), a spokesman for a bear who practises martial arts (d) and one who sounds like he knows his place in a church (e) are all found in a place where, on reflection and put crudely, there is nothing (f). How is this?

(3)

4. The following list of 55 words can be divided into 10 sets, all of different sizes. Put another way, there is one set of 10, one set of 9, etc. What word is in the set of one?

ABRIDGE	ARSENIC	AUTUMN	AVENUE	BAPTISE
BEAN	BEHIND	BORON	CARD	CONSTRICT
COYOTE	CUCKOLDS	DAWN	DEFEATED	DOG
EINSTEINIUM	ENAMOURED	FOLLOWING	FREQUENTLY	HAWAII
HERBIVOROUS	IMP	INTRAMUSCULAR	JOHNSON	LATER
LIE	MAN	METEOROLOGY	MOROCCO	NANOTESLA
NEXT	NIPPED	ODD	PARTY	POST
PURSUING	SCARE	SHINE	SMOOTH	SPAIN
SPEECH	STUBBS	SUBSEQUENT	SUCCEEDING	TECHNETIUM
TEXTER	THERMO	THIEF	THOROUGHBRED	TITANIUM
WAR	WINNER	WOBURN	WOLF	WORLD

(5)

5. Which of these words is the odd one out?

BAP, DEW, LOP, OIK, RED, SAW, SEW, WAS, WED

(1)

6. Which number comes next in this sequence:

-, -, 4, 3, 6, 6, 3, 13, 22, ?

(1)

7. Which word is the odd one out in each case?

(a) ANGKOR, BI, CHEW, GLASTONBURY, KO, PAPA, PAD, STOCK

(b) BET, HOLLIDAY, JAW, MENTOR, NET, STOCK, TEEM, TYLER

(1, 1)

8. What matrix ends this sequence?

$$\begin{pmatrix} \text{MAKE QUIZ} \\ \text{VERY HARD} \end{pmatrix} \begin{pmatrix} \text{PLUG USER} \\ \text{BITE EXCO} \end{pmatrix} \begin{pmatrix} \text{OPEN APSE} \\ \text{JUST OMIT} \end{pmatrix} \begin{pmatrix} \text{TOPE FOPS} \\ \text{TAPS TOWS} \end{pmatrix} \begin{pmatrix} ? \ ? \\ ? \ ? \end{pmatrix}$$

(2)

9. Which element is missing?

ARSENIC, ASTATINE, BISMUTH, CARBON, COPPER, IRON, KRYPTON, NEON, OGANESSON, PHOSPHORUS, SILICON, TENNESSINE, TIN, XENON

(1)

10. Number search . . .

```
N  E  E  T  F  I  F  S  N
O  B  U  T  P  E  W  I  I
I  H  E  Z  E  N  I  X  N
L  C  H  E  E  I  N  T  E
L  U  R  R  R  N  M  E  T
I  B  O  O  H  H  E  E  H
M  F  O  R  T  Y  T  N  O
N  R  T  D  S  O  E  S  U
E  T  T  H  U  E  X  I  S
V  S  W  E  N  A  R  C  A
E  H  O  R  I  E  T  E  N
L  N  V  W  M  E  A  L  D
E  I  G  H  T  Y  O  N  E
```

(2)

11. The members of each list below are given in a particular order. Place them in a different but related order.

   (a) ARGENTINA, MONGOLIA, DOMINICAN REPUBLIC, MALAYSIA, PAPUA NEW GUINEA, BENIN, CAMBODIA

   (b) LOUISA, HANS, BARBARA, ARTHUR, ERLE, GABRIEL, EDGAR, ALEXANDER, ROBERT, HARRIET

   (c) TIME, JACK, JESUS, LIVE, FOOT, RAG, FOOL, LEEK, HELL

   (d) DAVID, HARVEY, JAMIE, JERRY, JOHN, JONNY, JUSTIN, KUAN, LYNDA, TIM, TOMMY, VAN

   (1, 1, 1, 2)

12. Identify:

   (a) I, II, V

   (b) H, He, Li, Be, C, F, Mg, Ar

   (1, 1)

13. For each word in this list it is possible to change the first letter, then the second, then the third and then the fourth to make a new word each time. The final words in each case are related and in alphabetical order. What are they?

   LAND, SLOT, SHIP, SHOW, WILD, KILT, ITEM, WHAM, CHOP, HAVE

   (2)

14. Four 16-letter phrases have been enciphered by adding a key stream consisting of a repeating 4-letter word (A=1, B=2, etc.). The same key stream is used in each case. One of the letters in the 4-letter word is A. Two words are missing. What are they?

WUBQ	IQSE	HJTX	XJDU
VXFR	LUTU	HFTX	XJDU
MXFZ	XQOV	GWPS	EYGR
GECB	WOFK	IUDG	LJIR

(2)

15. Whic to erms bgn a f - l qu?

(a) ?, ?, GN, B, U, PIK, AC

(b) ?, ?, H, U, -, X, V, G, -, -, L, W, ...

(c) ?, ?, 2, 3, 46, 05, 7, 98, ...

(d) ?, ?, O, RM, TANG, -, F, D, -, P, L, -, -, X, -, -, J, -, -, Z, -, -, V, -, -, -, -

(e) ?, ?, EN, CHMD, O, Z, QT, W, -, P, -, F, -, K, -, G, -, -, -, -, -, -, -, -, -, -, -, -, -, -, -, -, -, -, X, -, J, ...

(f) ?, ?, GL, TOS, CN, H, X

(g) ?, ?, IE, FOR, -, H, QZ, -, ND, -

(1, 1, 1, 1, 1, 1, 1)

16. Which element?

Scandium, Titanium, Vanadium, Manganese, Iron, Cobalt, Copper, Zinc, Gallium, Arsenic, Selenium, Bromine, Zirconium, Technetium, Silver, Indium, Antimony, Iodine, Lutetium, Hafnium, Tantalum, Rhenium, Gold, Mercury, Thallium, Bismuth, Astatine

(1)

17. Webrb, orbpfsbiy, nfn mlb mc teb sbttbrs emifnjy tefs ybjr?

NYRIPTGY, NYRJGY, DPDJGJRI, DQNDLYCD, CJNYGY, RYCIRJGD

(2)

18. Which book? There's an encipherment clue in the alternative title.

ET UGWQHENXLW NEEKS, FXNJN FYKXVG, T KYISL UYGE, QZLK APRT NRK NALJV PPEL, JLXRWGW PG YVK QIIO, VNT UJ AAV JKYEQT-PEE QSXA HOX FBFR, DIMXR XFRLVVKV, NDVMH EGV JDWGIYL, OOVDSND SQF FF XOQQ, O XFFF QSDNSDV, OO LYI GWDJVGEH, MVDNVK VVLRFXRXMID, RFB FX OHAX, SQEE PTQ EYKJV, AAJ UTNVOIIQK VY GPW GWRXLJ, L EPPVJ LA NKR NTRPA, KLZ QEGYUL SIYKT, GVX KYFNJN XFSKEEN, WLX NEHQ WGUTE

(2)

19. The Setters have been exploring new markets and are writing a cookery book, extracts from which are below.

(a) What's the secret ingredient?

DPT MFUUVDF

QFCIHCBG

OHPRQ MXLFH

WTQDM WQT

TVV

SKNYAOPANODENA OWQYA

KUJLT YNYYNA

LFWQNH

ITKFXLTG VAXXLX

(b) What sort of cake can be made from these ingredients?

EGGS

LEMONS

PLUMS

PEARS

APPLES

ELDERBERRY

NUTMEG

ICE CREAM

PEACHES

(1, 1)

20.   Place in chronological order. *(Where encrypted, please give your answer unencrypted.)*

(a) CNZUJW EJMU NJMUA, FUCPT YYLTULF ERLZ, HJRLIR EATJWAJWEBFM, MJNRB XWIJCX AIN JHSCM, NJWHFUJ FFLXQFB YVFBRHFWC, RDEFO EIDUXRJB MIFT, XRPTJN-VNF-YQUI YWHMRUNFM, XXSPE FBV J KFKJWT, XXUMTCULL OKBURBJM, XXUUE FDA JR ENHRQB

(b) AFYGVMCI, CFCI, GJMYI, KIWP, PIJAI, SIKGIJK, VTQDK, YTPMCK

(c) AGNOSTIC, ARCHDEACON, BEAUJOLAIS, BONFIRE, CATERPILLAR, GAUCHO, NIGHTJAR, SUBJUNCTIVE, SYMPATHY, TELEVISION, WATCHMAKER, WHISPER

(1, 1, 1)

21. The words before the colon all share a property. The word after the colon doesn't have that property, but has a similar, distinguishing property. Explain.

    (a) APPAL, CHIP, LAMB, MADE, MAT, MICRO, SILO, TAG, TAT, THE: PEAT

    (b) CARP, ERRATA, FOR, HIS, HOT, KEN, LIED, OSIER, RAY, TOR: ONE

    (c) AFLAME, DALE, FRESH, HAY, HINT, LAME, LET, MELD, SAD, THE: STEP

(1, 1, 1)

22. What's the next sequence:

    (i)   1, 2, 3, 4, 5, 7, 8, 9, 10, 11, ...
    (ii) 1, 2, 3, 4, 8, 10, 12, 13, 14, 15, ...
    (iii) 1, 2, 3, 4, 5, 7, 8, 9, 10, 11, ...
    (iv) 1, 2, 3, 4, 5, 13, 14, 15, 21, 22, ...
    (v) 1, 3, 4, 5, 6, 7, 8, 9, 10, 11, ...
    (vi) 5, 6, 7, 8, 9, 13, 15, 16, 17, 18, ...
    (vii) 1, 3, 5, 6, 7, 8, 9, 10, 11, 12, ...
    (viii) 1, 2, 3, 5, 6, 7, 8, 9, 10, 11, ...
    (ix) 1, 3, 5, 6, 7, 8, 9, 10, 11, 12, ...
    (x) ?

(1)

23. What is missing from this sequence?

SUM, SUN, SIX, –, SIP

(1)

24. Whose works? *(Any spaces and punctuation have been removed and the lists – not necessarily complete – were originally in alphabetical order; '-' denotes a list member of 0 letters.)*

   (a) TTWT, BBYG, B, VPP, BY, GTXPTT, T, TTT, TZZWT, BY, VTWT, T, TYTYW, TTYP, TPWPP (7, 4, 6, 7)

   (b) HMPFO, OH, PC, PC, O, HOO, HMKO, HPOFPZC, HOC, HYOMOFH, HP, YJY (7, 3, 8)

   (c) MMU, GNNN, MN, NBYNW, Y, WNW, B, BUN, NGNN, UN, B, MNWNWMU, UBWY, WNGMN, -, Z, NUN, UNN, VG (6, 6, 9)

   (d) ADW, A, AABY, CM, FCY, MM, AVW, KA, D, UFM, FMA, AM (3, 7, 6)

                                                    (1, 1, 1, 2)

25. Pair:

   (a) BELARUS, BENIN, DEER, DEFICITS, DEMURE, ERNEST, ESTONIA, FIJI, GREECE, HAITI, HOTTEST, INDIA, IRAN, MERITED, NEPAL, NIUE, SERENE, STEP, TIFF, USAGE

   (b) AGAMA, DOWN, GETAWAY, GNOME, INDIA, INTAKE, ISLE, MALTA, MORMON, NIGER, OMAN, OTHER, PERU, RAFT, SAMOA, SERBIA, SIX, TOGO, UKRAINE, YEMEN

   (c) BEHAVE, CRY, DENIZEN, DETAIL, DEVOUT, DIAL, ERITREA, GAMBIA, INDIA, IRAN, LATVIA, MALTA, NIUE, NOOSE, REST, RUSSIA, SAMOA, SUDAN, WEIGH, WIDE

                                                    (1, 1, 1)

26. Where does 2 fit in the following lists? *(Read left to right, top to bottom.)*

(a)

17	18	23	19	11	27	53	15	43	25	47
14	28	33	37	13	3	29	16	12	21	7
55	51	45	48	35	49	5	41	24	38	8
39	31	9	26	1	22	10	20	54	56	52
50	44	34	46	42	40	4	36	32	30	6

(b)

29	9	24	23	7	20	22	21	3	30	31
1	10	12	14	13	11	33	32	43	39	38
44	34	42	40	41	37	36	35	28	27	26
25	4	8	6	19	18	55	56	5	17	16
15	45	53	50	49	54	46	52	51	48	47

(c) (changing theme)

33	30	53	35	17	9	25	28	41	20	46
23	37	21	49	45	11	3	31	4	48	24
32	22	44	34	13	40	56	39	52	55	12
19	43	38	42	1	7	8	50	6	14	10
18	54	26	5	36	29	47	16	15	27	51

(2, 2, 2)

27. In a connecting wall you have to partition sixteen items into four connected sets of four – such a partition is called a solution. This connecting wall has 5 solutions (some simpler to spot than others!) To help you out no item appears in the same set as another item in any of the solutions more than once. What is the missing item?

25×1	1×3	0+8	11+11
1024+11	1056.0/16	16×31	19DA−57
121−94	???	117.0−111	925−259
35.0+806	9801/1089	5D091/1111	14020800/1725

(3)

28. (a) Who are 2B/37 better known as?

(b) Who 'composed' $91\pi/216000$?

(1, 2)

29. You wake up in the marvellous AL. Later you are joined by a DLIXVIZXH, a MRG MZNWLLD and a MLRO. Who should you all go off to see? (Please encrypt your 6-letter answer).

(1)

30. Which letter does this sequence suggest?

11, 33, 11, 25, 50, 25, 6, 18, 6, 12, 26, 8, 36, 2, 40, 10, 24, 18, 28, 22, 16, 32, 14, 38, 6, 34, 4

(1)

31. To what do these numbers lead you?

-, 3, 1, (5,6), -, -, -, -, 2, -, ...

7, 2, -, 2, -, -, 7, 5, 8, 2, ...

-, -, -, -, -, 2, -, -, -, -, ...

3, -, -, -, -, -, -, -, -, -, ...

(1)

32. What is the last word/Ce qui est le dernier mot:

(a) USE, FIX, QUIDS, CURT, EXIT, TUN, NORTH, EQUIP, AXIS, ?

(b) EN, VOIE, TWIST, ENFERS, HIER, ONT, HUNE, VEXE, TOGE, ?

(1, 1)

33. The Kwiz archive is full of partially formed ideas for kwestions for which even the Setters can't remember the details. For example, we found the following sentence in the archives, but there was clearly something missing. All we could remember was that what was missing was a number. The missing number is shown by a # in the sentence – but which number is it?

Before the quick sepia coloured fox jumped over the more lazy dog it saw # tapirs waiting in a nearby village.

(2)

34. Now some alkemy. Which is incorrect?

(a) ALUMINIUM + EINSTEINIUM = PALLADIUM

(b) BROMINE + URANIUM = CHLORINE

(c) COBALT + GOLD = NICKEL

(d) DYSPROSIUM + GOLD = OSMIUM

(e) EINSTEINIUM + COPERNICIUM = RUTHERFORDIUM

(f) FRANCIUM + BERKELIUM = RUBIDIUM

(g) GALLIUM + TUNGSTEN = IRON

(h) HELIUM + SILVER = RUTHENIUM

(i) IRON + SILVER = PLUTONIUM

(1)

35. What follows:

20, 18, 72, 52, 78, 60, 150, 30, 34, 51, ?

(1)

36. Where would you find?

(a) AGE, ATE, FOR, GIN, INN, LAW, ODE, OIL, ONE, OUT, RIM, TIC

(b) ASH, CAY, ERG, HAM, HUT, KIN, MAN, MAR, OUT, RID, ROE, TIC

(1, 2)

37. What is the next number in each of the following sequences:

(a) 13, 3, 45, 23, 24, 2, 24, 12, 24, 2, ?

(b) 3, 1, 3, 2, 25, 1, 12, 1, 3, 257, ?

(1, 1)

38. Where does HIPPOPOTAMUS fit?

(a) LION, CAT, RAM, BULL, PIG, HARE, HORSE, ELEPHANT, IBEX, DONKEY, GIRAFFE, ORYX, GAZELLE, JACKAL

(b) JACKAL, CAT, GAZELLE, ELEPHANT, GIRAFFE, HARE, RAM, IBEX, BULL, DONKEY, HORSE, PIG, LION, ORYX

(c) BULL, HORSE, DONKEY, RAM, PIG, CAT, JACKAL, LION, ELEPHANT, GIRAFFE, ORYX, GAZELLE, IBEX, HARE

(1, 1, 2)

39.

BRITAIN	-3 -1 (1,2) 1- 2- 2-	SPAIN
ECUADOR	-5 -4 -2 (1,2) (3,4) (1,4) 1+	PERU
GERMANY	-7 -4 (2,3) (3,4) (4,5) +5 1- 5+ 5+	FRANCE
GREECE	-6 -5 (1,3) (2,3) 2+ 2+	?
?	-2 (1,3) (3,4) 4- 4-	IRAQ
?	-8 -7 (2,5) 6-	?

(2)

40. One of the Setters accidentally merged and sorted columns E, F and G of his spreadsheet, giving a single column which began:

1, 1, 1, 2, 2, 2, 3, 3, 4, 4, 4, 5, 5, 6, 6, 7, 8, 8, 8, 9, 10, 10, 10, 11, 11, 12, 12, 14, 15, 17, 19, 20, 22, 40, 46, 60, 60, 60, 61, 64, 80, 80, 84, ...

What were the largest numbers in the original columns?

(2)

41. An odd kwestion with a message.

TOGS	HOCK	EACH	MODE	INCH
SUBS	SOAP	IMAM	NEAT	GODS
LADY	YEEL	RUDE	ITCH	CUBE
SIGN	FIGS	RUES	OMEN	MIFF
THEM	HIDE	ETCH	SWAN	EDGE
TUGS	TRAP	ENDS	RUBY	SOCK

(1)

Total (100)

Hints

Many of the puzzles in this book may look more intractable than they really are. In a lot of cases the trick is to approach the puzzle with the right mindset – in other words to think like a GCHQ puzzle setter. So for example when looking at a set of words, don't think of them as words, think of them as a set of letters. And when looking at a set of letters, think whether they could be converted into numbers. And when looking at a set of numbers, think whether they could be converted into letters.

Here are some examples:

What do the members of each set below have in common?

1. BOLIVIA, FRAGILE, JACUZZI, PICCOLO, VANUATU

2. ESTIMATE, ILLINOIS, PITILESS, STRUGGLE

What is the final member of these sequences?

3. D, H, L, P, T, ?

4. 1, 14, 19, 23, 5, ?

In question 1 there isn't really anything Bolivia and a piccolo have in common. But if you look at the words as a set of letters you will see they all have seven letters. That is something they have in common, but it's not particularly unusual. But if you look more closely you will see that each word ends in a vowel. That's a bit more unusual, and is on the right lines. Having spotted this about the last letters, have a look at the first letters – and you will notice that in each case the first letter of the word comes immediately after the last letter of the word in the alphabet. That is definitely unusual, and now you have the complete answer.

In question 2, again there isn't really anything Illinois and pitiless have in common. So look closely at the words – they all have eight letters. But what else have they in common? There is nothing particularly special about the first letters E, I, P, S – or the last letters E, S, S, E. How about every other letter – ETMT, ILNI, PTLS, SRGL? Again nothing particularly special there. So try looking inside the words. Do they contain hidden words? Well, ESTIMATE contains TIM and MAT, and ILLINOIS contains ILL, LIN and LINO. PITILESS contains several shorter words, but STRUGGLE only really contains RUG. Now you see the connection: the words contain MAT, LINO, TILES and RUG – all things you can put on a floor.

For questions 3 and 4, the letters in 3 don't make a word, and the numbers in 4 don't seem to be a numerical sequence. One thing to try is to convert letters to numbers and vice versa based on where they are in the alphabet. By this method A=1, B=2, C=3, D=4, E=5, F=6, G=7, H=8, I=9, J=10, K=11, L=12, M=13, N=14, O=15, P=16, Q=17, R=18, S=19, T=20, U=21, V=22, W=23, X=24, Y=25 and Z=26.

Using this conversion on the questions turns them into:

> 3. 4, 8, 12, 16, 20, ?

and

> 4. A, N, S, W, E, ?

Now the answers are clear. Question 3 is every 4th letter in the alphabet, so X is the final member. And question 4 spells out ANSWER, so the final member is 18, representing the R.

This sort of approach will help with many of the questions. Look at first letters, last letters, central letters, every other letter. See if you can add a letter to the front, or at the end, or in the middle, or if you can remove a letter somewhere. Look for hidden words, or hidden words in reverse. Look for the letters of the alphabet running through a sequence.

When converting letters to numbers and vice versa, there are ways other than A=1, B=2, etc. Letters could be converted into their Scrabble score (see Appendix), or possibly via their positions on a typewriter keyboard or mobile phone keypad (ABC=2, DEF=3, GHI=4, JKL=5, MNO=6, PQRS=7, TUV=8, WXYZ=9).

Letters may need to be interpreted as their NATO phonetic alphabet equivalents: A=ALFA, B=BRAVO, etc. (see Appendix).

Some number puzzles in the book may be mathematical, but they may also be based on how the numbers are written as words (ONE, TWO, THREE, etc.). The British method of writing numbers is used in this book so 108 would be written ONE HUNDRED AND EIGHT. Numbers may also refer to elements in the periodic table, or their symbols (see Appendix).

As well as English we use some French and German, particularly the numbers:

> French: UN, DEUX, TROIS, QUATRE, CINQ, SIX, SEPT, HUIT, NEUF, DIX, ...
>
> German: EINS, ZWEI, DREI, VIER, FÜNF, SECHS, SIEBEN, ACHT, NEUN, ZEHN, ...
>
> And we also use Roman numerals: I, II, III, IV, V, VI, VII, VIII, IX, X, ...
> (I=1, V=5, X=10, L=50, C=100, D=500, M=1000)

There are a number of common themes which run through this book, and having these in mind will help with many of the questions. Some of these – the Periodic Table, US states and US presidents, NATO phonetic alphabet, the Braille alphabet, the Greek alphabet, Morse code and Scrabble values – are listed in the Appendix. We are also fond of some shorter sets:

> Cardinal Points: North, East, South, West
>
> Rainbow: Red, Orange, Yellow, Green, Blue, Indigo, Violet
>
> Tonic Sol-fa Scale: Do, Re, Mi, Fa, So, La, Ti
>
> The 12 Days of Christmas: 12 Drummers drumming, 11 Pipers piping, 10 Lords a-leaping, 9 Ladies dancing, 8 Maids a-milking, 7 Swans a-swimming, 6 Geese a-laying, 5 Gold rings, 4 Calling birds, 3 French hens, 2 Turtle doves and a Partridge in a pear tree
>
> Nine Muses: Calliope, Clio, Erato, Euterpe, Melpomene, Polyhymnia, Terpsichore, Thalia, Urania
>
> Snow White's Seven Dwarfs: Bashful, Doc, Dopey, Grumpy, Happy, Sleepy, Sneezy
>
> Planets: Mercury, Venus, Earth, Mars, Jupiter, Saturn, Uranus, Neptune (and, until recently, Pluto)
>
> Zodiac: Aries, Taurus, Gemini, Cancer, Leo, Virgo, Libra, Scorpio, Sagittarius, Capricorn, Aquarius, Pisces
>
> Chinese Zodiac: Rat, Ox, Tiger, Rabbit, Dragon, Snake, Horse, Goat, Monkey, Rooster, Dog, Pig
>
> Mohs Scale of Hardness: Talc (1), Gypsum (2), Calcite (3), Fluorite (4), Apatite (5), Orthoclase feldspar (6), Quartz (7), Topaz (8), Corundum (9), Diamond (10)

There are also larger sets we like: Olympic host cities, London tube lines, countries and capital cities, books of the Bible, symphonies and their composers; days of the week, months of the year, and animals may make an appearance too.

You may be able to tell something about our literary and musical tastes from the questions. We read Shakespeare, Tolkien, A. A. Milne, Enid Blyton's Famous Five books, J. K. Rowling's Harry Potter books and, of course, Conan Doyle's Sherlock Holmes series. We watch James Bond films, Dr Who and Monty Python. We listen to the Beatles and the Rolling Stones.

In our leisure time we play Scrabble, Monopoly and Cluedo, but also darts and snooker. So as well as Scrabble scores for letters you may come across Monopoly properties, Cluedo characters, weapons and rooms; numbers around a dartboard; and the values of snooker balls.

There are a few phrases we like, and one particular word:

> To be or not to be, that is the question
>
> If music be the food of love, play on
>
> The quick brown fox jumps over the lazy dog
>
> Supercalifragilisticexpialidocious

Word origins (i.e. which language they come from) and rhyming words come up too.

We also like writing words as tables and reading other words from them. So for example:

5. What series do the following words convey: DORIC, IDEAL, TIMID, ALOFT, ESSAY?

As nothing stands out try putting the words in a table:

D	O	R	I	C
I	D	E	A	L
T	I	M	I	D
A	L	O	F	T
E	S	S	A	Y

If you then read letters horizontally and vertically, starting top left, you can find DO, RE, MI, FA, SO, LA, TI – with TI positioned so that DO can be read off again next:

D	**O**	**R**	I	C
I	D	**E**	A	L
T	I	**M**	**I**	D
A	**L**	**O**	**F**	T
E	S	**S**	**A**	Y

There are also word squares where the same words can be read across and down.

As you would expect from a GCHQ puzzle book there are questions which involve breaking codes. By far the most common is simple substitution cipher. In this method each letter in a message is replaced by another particular letter. You have to work out which letter is replaced by which, and this could be done by doing a frequency count of the letters in a message (you would expect the commonest letters to be those representing E, T, A, O, I, N . . .) or by looking for common words (A, THE, AND, etc.). Most substitution ciphers rely on a keyword or keyphrase which can be discovered once you know what the substitution alphabet is. This is most easily explained with an example:

Plain text: WHAT MIGHT YOU TURN TO WHEN YOU GET STUCK ON A PROBLEM

Cipher text: WRST GBORT YJU TUNI TJ WRPI YJU OPT QTUMD JI S KNJEFPG

The plain text has been enciphered using this substitution alphabet:

Plain alphabet: ABCDEFGHIJKLMNOPQRSTUVWXYZ

Cipher alphabet: SEMAPHORBCDFGIJKLNQTUVWXYZ

The keyword here is SEMAPHORE. This has been written first (apart from the second E) followed by the rest of the alphabet in order. This method can mean that letters towards the end of the alphabet are enciphered to themselves (TUVWXYZ in this case), which can be a clue to solving such ciphers.

As you would also expect from a GCHQ puzzle book there are some mathematical puzzles – but only recreational maths knowledge should be needed. Some series used are:

Triangular numbers: 1, 3, 6, 10, 15, 21, etc.

Square numbers: 1, 4, 9, 16, 25, 36, etc.

Prime numbers: 2, 3, 5, 7, 11, 13, 17, 19, 23, etc.

Powers of two: 1, 2, 4, 8, 16, 32, 64, 128, etc.

Fibonacci sequence: 1, 1, 2, 3, 5, 8, 13, 21, 34, etc. (each number is the sum of the previous two)

Pi (3.14159265358979 . . .) comes up a few times, and e (2.71828182845 . . .) makes an appearance too.

Sometimes different bases are used. Popular ones are:

Binary (base 2): 1, 10, 11, 100, 101, 110, 111, 1000, 1001, 1010, etc.

Hex (base 16): 1, 2, 3, 4, 5, 6, 7, 8, 9, A, B, C, D, E, F, 10, 11, 12, etc.

Other bases are sometimes used too.

Finally, as well as the Where? and Which? questions, which are explained on pages 145–9, there are some other regular questions:

A to Z: In these questions the letters A to Z before the '=' represent single-word members of a particular set of items which start with these letters, and the letters after the '=' represent the initials of people or things which form all or part of the member of the set.

Chains: In these questions a chain can be built up from pairs of words or names which overlap in some way. You have the first and last members of the chain and have to work out the missing links.

Sums: In Sums questions you are presented with a strange-looking sum. Sometimes it consists of words, and sometime pictures. For these you need to find a number associated with each word or picture – and these associations will be of the same kind in each question. You need to identify the numbers and then calculate the answer to the sum. The answer will also have a number associated with it in the same way, and the answer to the question will be a word or picture of the same kind as those in the question.

For example:

(evens × ether – eon) / tow = ?

To solve this you need to spot the connection between all the words – which is that they are all anagrams of numbers. So using the associated numbers gives:

$(7 \times 3 - 1) / 2 = 10$

Ten is an anagram of net – so the answer to the question is 'net'.

Explanation of Puzzle Types

There are a couple of puzzles that appear regularly in this book but which need a little explanation. These are called 'Where?' and 'Which?'.

Where?

These questions consist of a list of words. Although written in columns for convenience, the words are just one list which should be read from left to right and from top to bottom. The question itself asks you where in the list an additional word should be placed.

The way the puzzle is constructed is that the words in the list can be divided into sets depending on words associated with them, and the way these words are associated. Each set is of the same size and has words associated in a different way, and these associated words are in alphabetical order through the list. Typically there are 7 sets of 8 words, or 8 sets of 7 words, though other arrangements may appear. The additional word belongs in one of the sets, and you have to work out where in the list the associated word fits alphabetically, and hence where the word itself should appear in the list.

An example may help. Here is a sample question:

> Where does JACKSON fit in the following list? *(Read left to right, top to bottom.)*

HUNT	WALL	SPROUTS	LAUGHING	LAUGHTER
NODDING	ROBINSON	EAST	RATHER	BUSH
NANNY	CLOTHES	KIRK	KENNEDY	PRIDE
BOTHER	HOLLY	HILTON	GUINEA	GRANT
CROSS	BLACK	BROOK	MISTER	WON
SYNDROME	CORONATION	DOC	PACT	

To solve this you need to find some sets. For example, you might see SPROUTS and PACT and think of BRUSSELS SPROUTS and WARSAW PACT. Brussels and Warsaw are European capitals, so you then look for more words which can be preceded by European capitals. This leads you to:

	Berlin	Brussels		
HUNT	WALL	SPROUTS	LAUGHING	LAUGHTER
NODDING	ROBINSON	EAST	RATHER	BUSH
				London
NANNY	CLOTHES	KIRK	KENNEDY	PRIDE
		Paris		
BOTHER	HOLLY	HILTON	GUINEA	GRANT
CROSS	BLACK	BROOK	MISTER	WON
Stockholm			Warsaw	
SYNDROME	CORONATION	DOC	PACT	

Note that the associated words, in this case the capital cities, are spread through the list in alphabetical order. They can now act as a guide to other sets.

Next you might notice CORONATION and EAST, and think of soap operas, or you might see NODDING and NANNY and think of farm animals. Filling in these sets gives:

	Berlin	Brussels	Cow	
HUNT	WALL	SPROUTS	LAUGHING	LAUGHTER
Donkey		Enders		
NODDING	ROBINSON	EAST	RATHER	BUSH
Goat	Horse			London
NANNY	CLOTHES	KIRK	KENNEDY	PRIDE
	oaks	Paris	Pig	
BOTHER	HOLLY	HILTON	GUINEA	GRANT
roads	Sheep	side		
CROSS	BLACK	BROOK	MISTER	WON
Stockholm	Street	tors	Warsaw	
SYNDROME	CORONATION	DOC	PACT	

The associated words have here been written above the words in the list for convenience – the sets are:

> LAUGHING Cow, NODDING Donkey, NANNY Goat, CLOTHES Horse, GUINEA Pig, BLACK Sheep

And:

> EASTenders, HOLLYoaks, CROSSroads, BROOKside, CORONATION Street, DOCtors.

Note that each set so far found has 6 members, so there must be 5 sets of 6 words – and JACKSON must belong to one of the missing sets.

Looking at the remaining words, you might notice that HUNT has to be associated with a word that comes before Berlin alphabetically. This might lead you to think of AUNT, which can be formed by changing the first letter. Similarly LAUGHTER could become DAUGHTER in the same way. Other words in this set are RATHER/FATHER, BOTHER/MOTHER, MISTER/SISTER and WON/SON. This gives you six words and six family members, so this set is also complete.

The remaining words are ROBINSON, BUSH, KIRK, KENNEDY, GRANT and of course JACKSON. These look like surnames of famous people, but there needs to be a link more specific than that. If you think of GEORGE W. BUSH and JOHN F. KENNEDY then you'll see the link is that these are people known by names including a middle initial. The others in the list are EDWARD G. ROBINSON, JAMES T. KIRK and RICHARD E. GRANT. JACKSON must belong to this group, and indeed SAMUEL L. JACKSON fits the pattern. To answer the question, you just need to say where JACKSON would go in the list, based on where SAMUEL fits alphabetically.

The complete list is therefore:

Aunt	Berlin	Brussels	Cow	Daughter
HUNT	WALL	SPROUTS	LAUGHING	LAUGHTER
Donkey	Edward G.	Enders	Father	George W.
NODDING	ROBINSON	EAST	RATHER	BUSH
Goat	Horse	James T.	John F.	London
NANNY	CLOTHES	KIRK	KENNEDY	PRIDE

Mother	oaks	Paris	Pig	Richard E.	
BOTHER	HOLLY	HILTON	GUINEA	GRANT	
roads	**Samuel L.**	Sheep	side	Sister	Son
CROSS	**JACKSON**	BLACK	BROOK	MISTER	WON
Stockholm	Street	tors	Warsaw		
SYNDROME	CORONATION	DOC	PACT		

So the answer is that JACKSON fits between CROSS and BLACK.

Which?

These questions consist of a list of 55 words, which can be divided into 10 sets, all of different sizes. Put another way there is one set of 10 words, one set of 9 words, one set of 8 words and so on down to one set of two words and one set of just one word. You need to identify the word in the set of one.

The way these questions are constructed is that the words in each set are connected to the size of the set. So the words in the set of 10 have some connection with 10, the words in the set of 9 have some connection with 9, and so on. Again an example might help – in this case cut down to just sets of sizes one to 5.

Which word is in the set of one?

ARAMIS	ARCTURUS	ATHOS	BRIAN	CANOPUS
DIVE	GIVE	HITHER	JIVE	LIVE
ON	PORTHOS	PROCYON	RIGEL	SIRIUS

In this case it may be that the names ATHOS, PORTHOS and ARAMIS stand out. These are The THREE Musketeers. You might spot some of the other names are those of stars – ARCTURUS, CANOPUS, PROCYON, RIGEL and SIRIUS. There are five of them, so it's a FIVE Star solution!

That leaves:

| BRIAN | DIVE | GIVE | HITHER | JIVE | LIVE | ON |

Four of these have four letters – but more to the point they have IV, the roman numeral for FOUR, in their centres. Of the remaining three words, both HITHER and ON form new words if suffixed by TO (HITHERTO, ONTO), which is a homophone for TWO. So the answer is BRIAN. This is associated with ONE, as a famous BRIAN is BRIAN ENO, and ENO is ONE reversed.

So the 5 sets are:

> 5 – Five Star: ARCTURUS, CANOPUS, PROCYON, RIGEL, SIRIUS
>
> 4 – IV: DIVE, GIVE, JIVE, LIVE
>
> 3 – Three Musketeers: ATHOS, PORTHOS, ARAMIS
>
> 2 – *to: HITHERto, ONto
>
> 1 – ENO: BRIAN

Clues to the Starter Puzzles

1. Think of the publisher

2. Think films

3. What star sign are you?

4. What letters do the answers have in common?

5. The first word is MOSES

6. Which countries are these the capitals of?

7. Read the answers backwards and forwards

8. Have you got double vision?

9. i. Think length; ii. Think Bungle, Dorothy or Noah; iii. It's all in the name

10. Think Roman synonyms

11. The answers are very similar

12. Look for words which could follow those in the question

13. Where would you START with this wine?

14. Look in the larder

15. Which planet is this German Count on?

16. Nudge nudge, wink wink, say no more

17. Think in colour

18. Look within the country names

19. Suit you, Sir

20. When in Rome

21. Numbers refer to positions of letters

22. After 39 steps . . .

23. Speak them out loud and go with the flow

24. One of the pairs is 8. Beckham and 12. Peckham

25. Try adding a letter to sing a good carol

26. Try jumbling the letters to get some familiar friends

27. Convert the letters to numbers

28. Convert the numbers to letters – and look at neighbouring letters

29. Can you see the wood for the trees?

30. Think phonetics (see Appendix)

31. (a) Think size; (b) Think team; (c) Think letters

32. These are initial letters of words in book titles

33. Think snooker

34. SOME questions don't need clues

35. The fourth letter is O

36. Look at the clock

37. Try waving a wand over the question

38. You don't have to be able to reed music to answer this question

39. Look within the words

40. Changing a NAME is MEAN

41. Read the initial letters and follow the instruction

42. Just kick this question around

43. Try browsing the internet

44.   Look at the vowels

45.   The first answer is EYE – hacknEYEd

46.   Question TIME

47.   All the questions, and the title, are anagrams

48.   Look for 2-letter synonyms

49.   Look at the Appendix

50.   Try writing the numbers as words

51.   Look for our most famous dramatist

52.   Think phonetics (see Appendix)

53.   Look for a shapely answer

54.   Gotta catch 'em all

55.   Try adding words, and if that doesn't help, ask your parents

56.   Think board game

57.   What words might follow the words in the question?

58.   Only the middle letters are the compass points

59.   Don't think in English

60.   Are you primed for this one?

61.   THIS maze may lead you down the wrong paths

62.   Think of a boy band

63.   Who travels through time?

64.   Think chess

65.   Try turning the triangle to find the quotation

66.   Think in colour

67.   What groups do the words belong to?

68.   What LINKS?

Answers

Section 1

1.

These are all species of **penguin**.

2.

Spock Sith. The sequence is formed from the last words in the titles of Star Trek/Star Wars films, in order. The third films in each series are The Search for Spock and Revenge of the Sith.

3.

Scorpion. The sign of the zodiac on each day determines what we eat:
St David's Day is 1 March – Pisces
US Independence Day is 4 July – Cancer
Christmas Day is 25 December – Capricorn
Halloween is 31 October – Scorpio

4.

    (a) **Cent**
    (b) **Kent**
    (c) **Lent**
    (d) **Tent**

They all end ENT, and Ents were created by J. R. R. Tolkien in The Lord of the Rings.

5.

MOSES, OFTEN, STARE, EERIE, SNEER

M O S E S
O F T E N
S T A R E
E E R I E
S N E E R

6.

Budapest. The European countries of which these are the capitals begin with A, B, C, ... Austria, Belgium, Czech Republic, Denmark, Estonia, Finland, Greece, Hungary.

7.

    (a) **KAYAK**
    (b) **MINIM**
    (c) **NOON**
    (d) **EVE**
    (e) **EWE**

These are all palindromes, as are AHA and REFER!

8.

Cormorant. This is hidden in the double letters: raCCOOn paRRot, leMMing, kOOkabuRRa, AArdvark, liNNet, oTTer

9.

i.   **5-letter** words / not five-letter words

ii. Colours of the **rainbow** / not colours of the rainbow

iii. Common **surnames** / not common surnames (with apologies to Rabbi Lionel Blue and Jason Orange)

10.

10015150. MINGLE is a synonym of MIX, which becomes 1000110 when each letter (all represent Roman numerals) is replaced by its decimal equivalent; similarly for APE/MIMIC/1000110001100. POLITE is a synonym of CIVIL, hence the answer.

11.

(a) **2020**

(b) 20/20

(c) **Twenty20**

12.

(a) **League**. This can precede Table; all the others can precede Chair.

(b) **Rose**. This can precede Garden; all the others can precede House.

(c) **Glass**. This can precede Ceiling; all the others can precede Floor.

13.

Tea, or any drink beginning with T. The initial letters spell MERLOT.

14.

    (a) **Bread**
    (b) **Brown**
    (c) **Mint**
    (d) **Tomato**
    (e) **Worcestershire**

These are all **sauces**.

15.

    (a) **N**. Initial letters of planets. N for NEPTUNE.
    (b) **E**. Initial letters of German numbers. E for ELF.

16.

Terry. These are the first names of the Monty Python Team. Terry is missing, but as there is another Terry, it is also not missing.

17.

    (a) Ultra VIOLET
    (b) Mood INDIGO
    (c) BLUE
    (d) GREENsleeves
    (e) YELLOWstone
    (f) ORANGEs and lemons
    (g) RED rum

You might find **gold** at the end of this rainbow.

18.

Gary is in **hunGARY**. His friends are in argenTINA, denMARK, RUSSia, suDAN and the uniTED states.

19.

13. Broken and Purple can be followed by Heart; Rough and Baseball can be followed by Diamond – these are red suits in a pack of cards. Sam and Garden can be followed by Spade; Golf and Fight can be followed by Club – these are black suits. Suits contain 13 cards.

20.

Mike. Each word is the Nato phonetic alphabet word for the Roman numeral equivalent to the number. Mike is the word for M, which is the Roman numeral for 1000.

21.

Beethoven's 9ᵗʰ. Taking the identified letter from each composer's name spells out BEETHOVEN.

22.

FORTY (winks!)

23.

They are British rivers: Dee, Exe, Forth, Wye.

24.

The pairs are:

1. Terry	**9.**	Kerry
2. Matalan	**11.**	Catalan
3. Panama	**7.**	Manama
4. Jersey	**5.**	Mersey
6. Mork	**10.**	York
8. Beckham	**12.**	Peckham

We hope you weren't put off by any of: (John) Terry/Kerry, (County) Cork/York, Nottingham/Mottingham or Lune/June!

25.

KING WENCESLAS.

SQUAW**K**, SCAMP**I**, ASTER**N**, BASIN**G**, HALLO**W**, CLOTH**E**,
RATIO**N**, PARSE**C**, HEARS**E**, CARES**S**, GRAVE**L**, CHORE**A**,
ASSES**S**

26.

HOOP. The set are Christopher Robin's animal friends with their
letters in alphabetical order: Kanga, Rabbit, Eeyore, Tigger, Piglet,
Owl, Roo. Missing is Pooh, who becomes HOOP.

27.

(a) **U**. Converting letters to numbers using A=1, B=2 gives
1,3,6,10,15 which are triangular numbers. The next is 21.
(b) **P**. Converting as above gives 1,2,4,8 which are powers of 2. The
next is 16.
(c) **W**. Converting as above gives 2,3,5,7,11,13,17,19 which are
prime numbers. The next is 23.

28.

4. Each number is the position in the alphabet of the letter preceding
it: t(20), s(19), d(4), h(8), l(12), d(4), r(18), **d(4)**, a(1).

29.

FIRM. Each word contains a 3-letter tree: FIRm, hELM, sOAK,
wASH, but only in FIRM is the tree at the beginning.

30.

PAPA was a rolling stone; **ECHO** beach; **HOTEL** California,
WHISKEY in the jar; **ROMEO** and **JULIET**. The missing words are
all from the NATO phonetic alphabet.

31.

(a) **Table tennis**. The sequence is ball sports, with size of the ball decreasing.

(b) **Field hockey,** or any other sports with 11 players in a team. The number of players in a team for each sport is 2,3,5,7, which are prime numbers. The next is 11.

(c) **Swimming,** or any other sport beginning with S. The initial letters spell out ATHLETICS.

32.

(a) **Charles Dickens**: Oliver Twist, A Christmas Carol, The Old Curiosity Shop; A Tale of Two Cities

(b) **Arthur Conan Doyle**: A Study in Scarlet, The Sign of Four, The Hound of the Baskervilles, The Valley of Fear

(c) **Evelyn Waugh**: Scoop, Vile Bodies, Decline and Fall, A Handful of Dust

33.

Maximum break. $7 \times (6+5+4+3+2+1) = 147 =$ maximum break in snooker.

34.

SOME. The other words can precede or follow it to become words:

AWESOME, FEARSOME, HANDSOME, LONESOME, TIRESOME, WHOLESOME, WINSOME

SOMEBODY, SOMEDAY, SOMEHOW, SOMEONE, SOMETHING, SOMEWHAT, SOMEWHERE

35.

IDIOT.

36.

1029. 3:55 = FIVE TO FOUR = 524; 1235 = TWENTY FIVE TO ONE = 2521; 850 = TEN TO NINE = 1029.

37.

Harry Potter and the. The letters are the initial letters of the ends of Harry Potter book titles: 1. Philosopher's Stone, 2. Chamber of Secrets, 3. Prisoner of Azkaban, 4. Goblet of Fire, 5. Order of the Phoenix, 6. Half-Blood Prince, 7. Deathly Hallows.

38.

Reed. Clarinets and Saxophones have a single reed; Bassoons and Oboes have a double reed.

39.

10. The numbers can be found inside the words. lONELy, neTWOrk, threaTENing.

40.

They are all anagrams of other names:

ALICE, CICELY, ELAINE, ESTHER, JANE, JASON, MARIAN, NORMA, RONALD

CELIA, CECILY, AILEEN, HESTER, JEAN, JONAS, MARINA, ROMAN, ROLAND

41.

NOEL. Reading left to right, the first letters of each item spell out PERFORM FREQUENCY COUNT. If a frequency count of the letters in the items in the list is performed it will be found that all letters are present save L, that is, there's 'no L'.

42.

A. The pairs are Premier League clubs and their grounds: Chelsea=Stamford Bridge, Crystal Palace=Selhurst Park, Everton=Goodison Park, Manchester United=Old Trafford, Tottenham Hotspur=White Hart Lane and Liverpool=Anfield.

43.

    (a) **Chrome**
    (b) **Edge**
    (c) **Opera**
    (d) **Safari**

These are all **web browsers**.

44.

Madagascar. The countries contain 1, 2, 3 and 4 identical vowels: cyprUs swEdEn, mOrOccO, mAdAgAscAr.

45.

hacknEYEd, mARMalade, bewitCHINg, arcHIPelago, battLEGround, astoniSHINg, potaTOEs, mercHANDise, freewHEELing, childbEARing, horRIBle, deLIVERance, naNOSEcond, orCHESTra, arGUMent, criSPINEss, thANKLEss, pLUNGer.

46.

    (a) **S.** These are the initials of every other day: Monday, Wednesday, Friday, Sunday, Tuesday, Thursday and Saturday.
    (b) **AN.** These are the first and last letters of the seasons: Winter, Spring, Summer and Autumn.
    (c) **D.** These are the initials of the last six months of the year: July, August, September, October, November and December.
    (d) **1.** 7 o'clock plus 8 hours is 3 o'clock. 10 o'clock plus 9 hours is 7 o'clock. 11 o'clock plus 2 hours is 1 o'clock.

47.

'Metro access' is an anagram of SOCCER TEAMS.

    (a) **Real Madrid**
    (b) **Scunthorpe United**
    (c) **Barcelona**
    (d) **Manchester City**
    (e) **Middlesbrough**

48.

> **VOLKSWAGEN.** The words represent AM, BY, CL, DJ, EG, FX, HQ, IT, OZ, PS, RN, UK, **VW.** The 13 answers are in alphabetical order and use each letter of the alphabet precisely once.

49.

> **Scrabble.** The scores are the Scrabble scores of the game names.

50.

> **After 6.** The first line is numbers with three letters, the second line is numbers with four letters, the third line is numbers with five letters.

51.

> (a) **Michelangelo**
> (b) **Pierre Fermat**

> Take the central three letters from each word, so: unWILls, foLIAte, guMSHoe, weAKEst, reSPEct, scAREdy gives William Shakespeare.

52.

> **Liverpool.** This can be succeeded by a letter in the NATO phonetic alphabet; the others can be preceded by one. The sequence involves the first 5 letters: Alfa Romeo, Bravo Two Zero, Charlie Brown, Delta Force, Liverpool Echo.

53.

> **O.**

> The sequence represents shapes and their numbers of sides. 3=Triangle, 4=Square, 5=Pentagon, 6=Hexagon, 7=Heptagon, 8=Octagon.

54.

> **VENUS.** The names can be found in the names of the Pokémon whose number is given. Pokémon number 3 is VENUSAUR.

55.

MA and **PA**.

(MA)DAM,(MA)INLAND,(MA)KING,(MA)LADY,(MA) LICE,(MA)LINGER,(MA)SON,(MA)STIFF,(MA)TINS

(PA)DRONE,(PA)INTER,(PA)LACE,(PA)NICKED,(PA) RENTAL,(PA)ROLE,(PA)STING,(PA)USED,(PA)WING

56.

Lounge. These are the pairs of rooms joined by secret passages in Cluedo.

57.

They precede things you might find in an **orchestra**: ear **DRUM**, champagne **FLUTE**, shoe **HORN**, Bermuda **TRIANGLE**, ear **TRUMPET** and super **CONDUCTOR**.

58.

Northern Ireland.

Northern Ireland is West	of Scotland	
Scotland	is North	of England
England	is East	of Northern Ireland
Wales	is West	of England

59.

ELF. This is German, the rest are French!

60.

2017. They all had prime number ages, in a year which was also a prime number. This happens again in 2017, when they will be aged 17, 19, 47 and 53.

61.

The 'centre' of the maze is the **D** in the 8ᵗʰ row, 7ᵗʰ column. THIS ROUTE IS THE ONE WHICH REACHES THE END.

T U O R S I H T H I

E I U O R S I H I S

T S T H D E R A S R

H G E E R R I S I O

E N I S N O N E T U

O O R W G R G I S T

N H T A P W E H T E

E E A C H N D R O I

W R H S E E T R C S

H I C T H E C E N I

62.

Harry Styles, Niall Horan, Liam Payne, Zayn Malik, Louis Tomlinson. These are the original members of One Direction – a 1D name!

63.

7. The words appear in the surnames of actors who have played the Doctor in Dr Who. William HARTnell (1ˢᵗ Doctor), Patrick trOUGHTon (2ⁿᵈ Doctor), Jon perTWEE (3ʳᵈ Doctor), (Tom and Colin) BAKEr (4ᵗʰ and 6ᵗʰ Doctors), Sylvester mcCOY (7ᵗʰ Doctor).

64.

ARCH(bishop), **PLAN**(king), **WEE**(knight), **FROGS**(pawn). Each word can be followed by a chess piece.

65.

Starting at the lower left and moving across the diagonals gives 'IF MUSIC BE THE FOOD OF LO?E PLAY ??', which is completed by:

N

V O

66.

Northern. The colours of the lines are red, yellow, green, brown, blue, pink and black – which score 1,2,3,4,5,6 and 7 points respectively in snooker.

67.

There are 4 members of **groups of 4**, 5 members of **groups of 5** and 7 members of **groups of 7**.

4: c. DIAMONDS (Suits), h. MILT JACKSON (Modern jazz quartet), j. LUKE (Gospels), m. SPRING (Seasons)

5: a. BORODIN (The Five), g. MICHAEL JACKSON (Jackson 5), k. SMELL (Senses), n. SUPERIOR (Great Lakes), p. U (Vowels)

7: b. COLOSSUS (Seven Wonders), d. ENVY (Sins), e. GREEN (Rainbow), f. HAPPY (Dwarfs), i. KELVIN (SI Units), l. SOLON (Sages), o. SUNDAY (Days of the Week)

68.

What links these is '**LINKS**', or '**LYNX**'.

Section 2

1.

RESURRECTION (Mahler) fits between PENCIL (Lead) and CIRCLE (Matarese).

Aquitaine	Bald	Barium	Beethoven	Black	
PROGRESSION	COOT	MEAL	CHORAL	THUNDER	
Bourne	Britten	Bruckner	Carbon	Chancellor	
IDENTITY	SIMPLE	ROMANTIC	COPY	MANUSCRIPT	
Cobalt	Cool	Copper	Drunk	Dvorak	
BLUE	CUCUMBER	BEECH	LORD	ENGLISH	
Fit	Gemini	Gold	Good	Haydn	
FIDDLE	CONTENDERS	RUSH	GOLD	SURPRISE	
High	Holcroft	Icarus	Iron	Lead	**Mahler**
KITE	COVENANT	AGENDA	AGE	PENCIL	**RESURRECTION**
Matarese	Matlock	Mendelssohn	Mozart	Neon	
CIRCLE	PAPER	ITALIAN	PRAGUE	LIGHTING	
Nitrogen	Old	Osterman	Oxygen	Parsifal	
CYCLE	HILLS	WEEKEND	TENT	MOSAIC	
Platinum	Prokofiev	Red	Rhinemann	Scarlatti	
BLONDE	CLASSICAL	BEETROOT	EXCHANGE	INHERITANCE	
Schubert	Schumann	Silver	Sober	Tight	
TRAGIC	SPRING	SCREEN	JUDGE	DRUM	
Vaughan-Williams	White				
SEA	GHOST				

Themes are:

Similes: Bald (as a) COOT, Black (as) THUNDER, Cool (as a) CUCUMBER, Drunk (as a) LORD, Fit (as a) FIDDLE, Good (as) GOLD, High (as a) KITE, Old (as the) HILLS, Red (as a) BEETROOT, Sober (as a) JUDGE, Tight (as a) DRUM, White (as a) GHOST

Elements: Barium MEAL, Carbon COPY, Cobalt BLUE, Copper BEECH, Gold RUSH, Iron AGE, Lead PENCIL, Neon LIGHTING, Nitrogen CYCLE, Oxygen TENT, Platinum BLONDE, Silver SCREEN

Symphonies: Beethoven (CHORAL), Britten (SIMPLE), Bruckner (ROMANTIC), Dvorak (ENGLISH), Haydn (SURPRISE), **Mahler (RESURRECTION),** Mendelssohn (ITALIAN), Mozart (PRAGUE), Prokofiev (CLASSICAL), Schubert (TRAGIC), Schumann (SPRING), Vaughan-Williams (SEA)

Robert Ludlum novels: (The) Aquitaine PROGRESSION, (The) Bourne IDENTITY, (The) Chancellor MANUSCRIPT, (The) Gemini CONTENDERS, (The) Holcroft COVENANT, (The) Icarus AGENDA, (The) Matarese CIRCLE, (The) Matlock PAPER, (The) Osterman WEEKEND, (The) Parsifal MOSAIC, (The) Rhinemann EXCHANGE, (The) Scarlatti INHERITANCE

2.

QUADRATS. The words are anagrams of SUNDAY, MONDAY, ..., FRIDAY and SATURDAY with the Y changed to another letter.

3.

(a) **SCOWL** has an animal in the middle, not at the edges: BE<u>GGA</u>R, BO<u>XCA</u>R, DE<u>LIVE</u>R, HA<u>RDWARE</u>, LI<u>TIGATION</u>, MO<u>LECU</u>LE, S<u>COW</u>L, SE<u>VERA</u>L, VO<u>LATILE</u>.

(b) **BAIT** can be preceded by WHITE, all the others by BLACK.

(c) **TROUBLES** (with A=1, ..., Z=26) doesn't have letters alternating odd/even, but the others do: BODY/eoeo, HOTEL/ eoeoe, KNIFE/oeoeo, RELAXING/eoeoeoeo, STATELY/ oeoeoeo, TROUBLES/eeooeeoo, UNGRATEFUL/oeoeoeoeoe, WRINKLE/oeoeoeo.

(d) **IDEA** is the only word that isn't Arabic in origin.

(e) **IDEA**: all have rhyming synonyms, but IDEA's ends in
-OUGHT, not -ORT: CASTLE/FORT, DEATH/MORT,
HARBOUR/PORT, IDEA/THOUGHT, LITTLE/SHORT,
RECREATION/SPORT, TYPE/SORT, WRONG/TORT.

4.

(a) ACTS, CRIMPS, GIRTH, HALT, HEAL, ITEM, NOGS,
PATCHER, SNEAKS, STOP, STUN, TRAP, TSARS

#1	#2 = anagram of #1	#3 and #2	#4 = anagram of #3
GODS	DOGS	CATS	ACTS
VASE	SAVE	SCRIMP	CRIMPS
GROWN	WRONG	RIGHT	GIRTH
STAPLER	PLASTER	LATH	HALT
EARTHY	HEARTY	HALE	HEAL
DIET	TIDE	TIME	ITEM
CANED	DANCE	SONG	NOGS
SERVE	VERSE	CHAPTER	PATCHER
SADDLER	LADDERS	SNAKES	SNEAKS
SNAP	PANS	POTS	STOP
BLOTS	BOLTS	NUTS	STUN
PLACER	PARCEL	PART	TRAP
PRIESTS	STRIPES	STARS	TSARS

(b) ATHENS, BERLIN, BUDAPEST, KUALA LUMPUR,
MADRID, MOSCOW, PARIS, ROME, STOCKHOLM, THE
HAGUE, TOKYO

#1	#2 = synonym of #1	#3 = language of #2	#4 = capital of origin of #3
BIRDWATCHER	ORNITHOLOGIST	GREEK	ATHENS
BOMBING	BLITZ(KRIEG)	GERMAN	BERLIN
BEEF STEW	GOULASH	HUNGARIAN	BUDAPEST
DAGGER	KRIS	MALAY	KUALA LUMPUR
BROAD-BRIMMED HAT	FEDORA	SPANISH	MADRID
OPENNESS	GLASNOST	RUSSIAN	MOSCOW
DEAD END	CUL-DE-SAC	FRENCH	PARIS
OPEN PIE	PIZZA	ITALIAN	ROME
GRIEVANCE OFFICIAL	OMBUDSMAN	SWEDISH	STOCKHOLM
DRUMBEAT	BOOM	DUTCH	THE HAGUE
PAPER-FOLDING	ORIGAMI	JAPANESE	TOKYO

5.

3637 or **3435**. These are the number of letters in the words in the title of New Zealand's national anthems, 'God Defend New Zealand' and 'God Save The Queen'. The others are Australia: 'Advance Australia Fair'; USA: 'The Star-Spangled Banner'; Canada: 'O Canada'; UK: 'God Save The Queen'.

6.

(a) **PARSON**.

PRIEST is the only monosyllabic word

PADRE is the only non 6-letter word

CURATE is the only word not starting with P

PEAHEN is the only word that isn't a religious office

(b) **484**.

4 is the only number not divisible by 11

66 is the only non-square

121 is the only odd number

1936 is the only non-palindrome

7.

(a) Train**S**potting

(b) **T**he Nightmare Before Christmas

(c) **A**irplane

(d) Fo**R** Your Eyes Only

(e) Ja**W**s

(f) Lolit**A**

(g) Empi**R**e of the Sun

(h) Footloo**S**e

(i) **T**he Graduate

(j) T**H**e Sound of Music

(k) Froz**E**n

(l) **F**ight Club

(m) Cl**O**verfield

(n) Ame**R**ican Beauty

(o) ButCh Cassidy and the Sundance Kid

(p) PrEtty Woman

(q) BAdlands

(r) LaWrence of Arabia

(s) RAin Man

(t) MoonstrucK

(u) ThE Silence of the Lambs

(v) AlieN

(w) SexTape

The numbers on each poster pick out letters from the titles as indicated above, spelling out STAR WARS: THE FORCE AWAKENS.

8.

(4 12 7 15 9 8). If each number is replaced by the first letter in its name (e.g. 1 → O(NE), 2 → T(WO)), then each sequence spells out a word. The odd sequence out spells a word (RENNET) when each number is replaced by the last letter in its name.

9.

(a) **All words contain 5 consecutive letters**: CERTAINLY contains no consecutive letters.

(b) **All words can be spelled using only the top row of a UK typewriter**: FLASH uses only the second row.

(c) **In each word any two letters are a multiple of 3 apart in the alphabet**: (i.e. they are formed from one of ADGJMPSVY, BEHKNQTWZ or CFILORUX): in GASEOUS letters come from all three groups.

10.

They are the answers to cryptic anagram clues:

Altered images; jumble sale; change of pace; change of heart; times are a-changing; regime change; change of name; redirected mail; small change; Eton mess; mixed emotions; mixed up in something.

Another such phrase could be ASHEN KNOT, which would be 'shaken not stirred'.

11.

    (a) **COAT**. A, B, C, etc. can be added to the end of words. COAT-I

    (b) **AIL**. Words can be preceded by A, B, C, etc. to form new words. J-AIL

    (c) **USER**. The words can have A, B,C, etc. inserted in the middle. US(H)ER

12.

Each author has written a book (or in one case a poem) with a title containing a character's first name(s) but no last name. The initial letters of these characters' names form the question 'WHAT LINKS THEM' to which the answer is the previous sentence.

Richmal Crompton	Just William
Johanna Spyri	Heidi
Lewis Carroll	Alice's Adventures in Wonderland
Wilbert Awdry	Thomas the Tank Engine
Vladimir Nabokov	Lolita
Alexander Solzhenitsyn	One Day in the Life of Ivan Denisovich
Arthur Conan Doyle	Sir Nigel
Rudyard Kipling	Kim
Benjamin Disraeli	Sybil
Harriet Beecher Stowe	Uncle Tom's Cabin
Henry Wadsworth Longfellow	Hiawatha
Frederic William Farrar	Eric (or, Little by Little)
Joyce Lankester Brisley	The Milly-Molly-Mandy Storybook

13.

PIZZAZZ. There is only one Z and only 2 Blanks, so a word with 4 Zs is impossible to play.

14.

In the nth part, the nth and (n+1)[th] letters of each word are given:

(a) **O LI**ttle **TO**wn **OF BE**thlehem **HO**w **ST**ill **WE SE**e **TH**ee **LI**e

(b) d**EC**k t**HE** h**AL**ls w**IT**h b**OU**ghs o**F** h**OL**ly f**A** l**A** l**A** l**A** l**A** l**A** l**A** l**A** l**A**

(c) th**EY** sa**ID** th**ER**e'd be sn**OW** at ch**RI**stmas th**EY** sa**ID** th**ER**e'd be pe**AC**e on ea**RT**h

(d) whi**LE** she**PH**erds wat**CH**ed the**IR** flo**CK**s by nig**HT**

(e) litt**LE** donk**EY** litt**LE** donk**EY** on the dust**Y** road

(f) gaude**TE** gaude**TE** chris**TU**s est natus ex maria virgi**NE** gaude**TE**

(g) o christ**MA**s tree o christ**MA**s tree thy leaves are so unchan**GI**ng

(h) a spacema**N** came travell**IN**g on his ship from afar

(i) hey mr churchil**L** comes over here to say we're doing splendid**LY**

(j) o come all ye faithful joyful and triumphan**T**

15.

(a) **One letter can be changed to form the surname of a US president**: <u>H</u>ARRISON, PO<u>L</u>K, <u>P</u>IERCE, G<u>R</u>ANT, HA<u>Y</u>ES, TA<u>F</u>T, HOO<u>V</u>ER

(b) **Adding one S to each word forms a plural, adding a second S forms a non-plural**: BRASS, CARESS, MILLIONAIRESS, NEEDLESS, PRINCESS, TIMELINESS

(c) **All are words when converted to hexadecimal**: 10/**A**, 190/**BE**, 2766/**ACE**, 57005/**DEAD**, 11325150/**ACCEDE**, 14613198/ **DEFACE**, 16435934/**FACADE**

(d) **Anagrams of composers**: BUTCHERS/**SCHUBERT**, DIVER/ **VERDI**, HANDY/**HAYDN**, HARLEM/**MAHLER**, LARGE/ **ELGAR**

16.

GIRTH. The original words pair into anagrams of antonyms:

AFT/HINT = FAT/THIN, AFTERS/LOWERS = FASTER/
SLOWER, ALOFT/SKIN = FLOAT/SINK, FILE/HATED = LIFE/
DEATH, FUSTIER/HATRED = SURFEIT/DEARTH, GLARE/
MALLS = LARGE/SMALL, ISLE/THRUST = LIES/TRUTHS,
LAYER/TALE = EARLY/LATE, PEACH/READ = CHEAP/DEAR,
SCENTED/STANCE = DESCENT/ASCENT, SERVE/SPORE =
VERSE/PROSE

with FELT and GROWN remaining. With GIRTH added the
pairings, FELT/GIRTH = LEFT/RIGHT, GROWN/GIRTH =
WRONG/RIGHT can be made.

17.

(a) A A Milne

(b) B B gun

(c) e e cummings

(d) H H Holmes

(e) I I who have nothing

(f) J J Abrams

(g) L L Cool J

(h) O O gauge

(i) S S Great Britain

(j) Z Z Top

18.

**WHITE (Snow) fits between SNUFFLY (Sneezy) and LIE
(State).**

Arms	Aunt	Avenue	Bashful	Bay
UP	SALLY	PENNSYLVANIA	SHY	HUDSON

Boots	C	Carp	Close	Cod
PUSS	MIDDLE	ENTER	GLENN	PIECE

Cousin	Crescent	D	Dab	Day
GERMAN	FERTILE	THREE	HAND	BIRTH
Doc	Dopey	E	Father	Goldfish
SURGEON	STUPID	VITAMIN	CHRISTMAS	BOWL
Grandfather	Grandma	Grumpy	H	Happy
CLOCK	MOSES	SURLY	CELLBLOCK	CHEERFUL
Hay	Herring	Jay	K	Lane
MAKE	BONE	BLUE	SPECIAL	PENNY
Line	Mackerel	May	Mother	Pike
FALL	SKY	MAGGIE	EARTH	LET
Ray	Road	Say	Secret	Sin
STING	ABBEY	FINAL	WORD	LIVE
Sister	Skate	Sleepy	Sneezy	**Snow**	State
SLEDGE	BOARD	TIRED	SNUFFLY	**WHITE**	LIE
Street	T	Tay	Terrace	Time
SESAME	MODEL	RIVER	FOOTBALL	JUST
Uncle	V	Venice	Way	X
SAM	HENRY	DEATH	HIGH	MALCOLM

Themes are:

X in Y: UP/*ARMS*, PUSS/*BOOTS*, FALL/*LINE*, WORD/*SECRET*, LIVE/*SIN*, LIE/*STATE*, JUST/*TIME*, DEATH/*VENICE*

Can be preceded by a relation: *AUNT* SALLY, *COUSIN*-GERMAN, *FATHER* CHRISTMAS, *GRANDFATHER* CLOCK, *GRANDMA* MOSES, *MOTHER* EARTH, *SISTER* SLEDGE, *UNCLE* SAM

Can be followed by a road type: PENNSYLVANIA *AVENUE*, GLENN *CLOSE*, FERTILE *CRESCENT*, PENNY *LANE*, ABBEY *ROAD*, SESAME *STREET*, FOOTBALL *TERRACE*, HIGH*WAY*

Synonyms of the names of the Seven Dwarfs: SHY/*BASHFUL*, SURGEON/*DOC*, STUPID/*DOPEY*, SURLY/*GRUMPY*, CHEERFUL/*HAPPY*, TIRED/*SLEEPY*, SNUFFLY/*SNEEZY* plus ***SNOW*** **WHITE**

Can be followed by a 3-letter word ending -AY: HUDSON *BAY*, BIRTH*DAY*, MAKE *HAY*, BLUE *JAY*, MAGGIE *MAY*, STING*RAY*, FINAL *SAY*, RIVER *TAY*, HIGH *WAY*

Can be followed by a single letter: MIDDLE *C*, THREE *D*, VITAMIN *E*, CELLBLOCK *H*, SPECIAL *K*, MODEL *T*, HENRY *V*, MALCOLM *X*

Can be preceded by a fish: *CARP*ENTER, *COD*PIECE, *DAB* HAND, *GOLDFISH* BOWL, *HERRING*BONE, *MACKEREL* SKY, *PIKE*LET, *SKATE*BOARD

19.

With a substitution alphabet of THEQUICKBROWNFXJMPSVLAZYDG the phrase in the nth part of the question was enciphered by advancing each letter by n letters within this alphabet. Thus the unenciphered question reads:

(i) WHEEL ON A UNICYCLE

(ii) SIDES OF AN ARGUMENT

(iii) MONTHS IN A QUARTER

(iv) LEAVES ON A FOUR-LEAF CLOVER

(v) CENTS IN A NICKEL

(vi) CANS IN A SIX-PACK

(vii) COLOURS OF THE ???????

Thus the missing word is the encipherment of RAINBOW, namely **MENVJPS**.

20.

(a) **CON**QUISTA**DOR**, **CR**OSSB**OW**, **DU**MBSTRU**CK**, **GUI**LDHA**LL**, **KI**LOBY**TE**, **LA**NDMA**RK**, **SHRUB**L**IKE**, **STAR**T**LING**, **WR**ITT**EN** have a **bird's name** in the outer letters: IN**TERN**AL has the bird's name internally!

(b) BEAT, CHAFE, CHIC, CLOVER, HIDEOUS, MODEL, STALK, TRIO, GARNISHED form words **when the last letter is advanced by one position in the alphabet**: SINGLE forms a word when the **first letter** is thus transformed.

(c) BRANDY, CRUISE, DECK, DOPE, EASEL, FROLIC, KIT, LANDSCAPE, SLEIGH, YACHT are derived from **Dutch**: CARNIVAL is derived from **Italian**.

(d) **ASTE**ROID, BR**EWST**ER, BUCK**THORN**, COMPEN**SATE**, E**NTHRO**NE, **OUTSHINE**, **STEW**ARD, XAN**THOUS** all contain **anagrams** of one of NORTH, SOUTH, EAST and WEST: LIMPIDLY has **no letter in common** with any of them.

(e) ART, BUDS, COMPARE, LEASE, LOVELY, ROUGH, SHALL, SHORT, TEMPERATE, WINDS all appear in **Shakespeare's Sonnet 18** ('Shall I compare thee to a summer's day'): SUMMARY doesn't.

(f) **CANVAS**, **FLASKS**, FLOR**IST**, **IMAGO**, MA**INLAND**, MA**UV**E, OR**IG**IN, TE**EN**S, V**ARMINT**, **ZO**OMING become **US states when 2 letters are changed**: CONNECT**IONS** requires **3 letters** to be changed.

(g) CHAIN/CHINA, ENEMY/YEMEN, MOAN/OMAN, PLANE/ NEPAL, REGALIA/ALGERIA, REIGN/NIGER, SERIAL/ ISRAEL are all **anagrams of countries**: ANIMAL/MANILA is the **anagram of a capital city**.

(h) GRATE/GREAT, DISCRETE/DISCREET, HOSE/HOES, PRIDE/PRIED, RUDE/RUDE, SEER/SERE, WEAR/WARE all have a **homophonic anagram**: PRAISE doesn't.

21.

(a) **One** Flew Over the Cuckoo's Nest (Ken Kesey) – They're out there. Black boys in white suits up before me to commit sex acts in the hall and get it mopped up before I can catch them.

(b) A Tale of **Two** Cities (Charles Dickens) – It was the best of times, it was the worst of times . . .

(c) **Three** Men in a Boat (Jerome K. Jerome) – There were four of us: George, and William Samuel Harris, and myself, and Montmorency.

(d) The Sign of **Four** (Arthur Conan Doyle) – Sherlock Holmes took his bottle from the corner of the mantelpiece and his hypodermic syringe from its neat morocco case.

(e) Slaughterhouse **Five** (Kurt Vonnegut) – All this happened, more or less.

(f) Now We Are **Six** (A. A. Milne) – I have a house where I go when there's too many people.

(g) **Seven** Pillars of Wisdom (T. E. Lawrence) – Some of the evil of my tale may have been inherent in our circumstances.

(h) BUtterfield **Eight** (John O'Hara) – On this Sunday morning in May, this girl who later was to be the cause of a sensation in New York, awoke much too early for her night before.

(i) The **Nine** Tailors (Dorothy L. Sayers) – 'That's torn it!' said Lord Peter Wimsey.

(j) **Ten** Days That Shook the World (John Reed) – Towards the end of September 1917, an alien professor of Sociology visiting Russia came to see me in Petrograd. (Some versions of this book begin with 'Toward' rather than 'Towards'.)

22.

First and last letter of each word:

(a) Deck the halls with boughs of holly, fa la la la la la la la la ('Deck the Halls')

(b) Nowell, nowell, nowell, nowell, born is the king of Israel ('The First Nowell')

(c) The Christmas we get we deserve ('I Believe in Father Christmas' – Greg Lake)

(d) What will your daddy do when he sees your mamma kissing Santa Claus ('Merry Xmas Everybody' – Slade)

(e) Chestnuts roasting on an open fire ('The Christmas Song')

(f) They never let poor Rudolph join in any reindeer games ('Rudolph the Red-Nosed Reindeer')

(g) Ring out those bells tonight, Bethlehem, Bethlehem ('Little Donkey')

(h) Right against the forest fence, by Saint Agnes Fountain ('Good King Wenceslas')

(i) Sing choirs of angels, sing in exultation. Sing all ye citizens of heaven above ('O Come, All Ye Faithful')

(j) Thus spake the Seraph and forthwith appeared a shining throng ('While Shepherds Watched')

23.

(a) UNDERRUN, KANGAROO, COLLOQUY, MATADOR, OVERDRIVE, ATHLETICS, ELEVEN, LACERATE, PALATINE, POLICEMEN **rhyme with ONE, TWO, …, TEN**

(b) VISTA, STATION, IONOSPHERE, EREMITE, ITEM, TEMPER, PERCOLATOR, TORRENT, ENTIRETY, ETYMOLOGICAL, CALLING where **the last 3 letters of each word are the same as the first 3 of the next**

(c) ONLY, JUST, FAIR, CLEAR, EMPTY, DISCHARGE, FIRE, PASSION, SUFFERING, UNDERGOING where **each word can be a one-word description of its predecessor** (order can be reversed)

(d) EXCUSE, SECURE, COURSE, ROUSED, SOARED, ADORNS, RANDOM, DOMAIN, MANIOC, ACTION, NOTICE, INCITE, ELICIT where **each word has 5 letters in common with its successor** (order can be reversed)

(e) BEGINNING, DEPARTURE, FIGURES, RATES, PITY, POST, GNOMES, SORROWS, BRUISE, LEAKS are **synonyms of books of the Bible in order** (but not necessarily consecutive): GENESIS, EXODUS, NUMBERS, JUDGES, RUTH, JOB, PROVERBS, LAMENTATIONS (Old Testament), MARK, REVELATION (New Testament)

24.

(a)

#1	#2 = origin of #1	#3 can be preceded by #2
SHAWL	PERSIAN	CARPET
COURTESY	FRENCH	CHALK
TULIP	TURKISH	DELIGHT
CARGO	SPANISH	FLY
BANDIT	ITALIAN	GARDEN
COACH	HUNGARIAN	GOULASH
MOLASSES	PORTUGUESE	MAN-O'-WAR
MYTH	GREEK	ORTHODOX
EDUCATION	LATIN	QUARTER
CORACLE	WELSH	RAREBIT
MAMMOTH	RUSSIAN	ROULETTE
POODLE	GERMAN	SHEPHERD
DECK	DUTCH	UNCLE

(b)

#1	#2 = homophone of #1	#3 = anagram of #2
STARE	STAIR	ASTIR
BEACHES	BEECHES	BESEECH
PEAK	PIQUE	EQUIP
FOURTH	FORTH	FROTH
RHYME	RIME	MIRE
SALTER	PSALTER	PLASTER
CARROT	CARET	REACT
BOARD	BORED	ROBED
CHASED	CHASTE	SACHET
IDLES	IDOLS	SOLID
COARSE	COURSE	SOURCE
CHOIRS	QUIRES	SQUIRE
ASSENT	ASCENT	STANCE
REST	WREST	STREW
PAUSE	PAWS	SWAP/WASP

(c)

#1	#2 = German number within #1	#3 = #2 in French
UNSTUFFY	FÜNF	CINQ
HOWITZER	ZWEI	DEUX
HAZELNUT	ZEHN	DIX
MATCHBOX	ACHT	HUIT
PUNGENCY	NEUN	NEUF
RELATIVE	VIER	QUATRE
SENSIBLE	SIEBEN	SEPT
SWITCHES	SECHS	SIX
DEADLIER	DREI	TROIS
QUESTION	EINS	UN

(d)

#1	#2 = anagram of #1	#3 = homophone of #2
LATER	ALTER	ALTAR
DOUBLER	BOULDER	BOLDER
ZEBRA	BRAZE	BRAISE
ASHORE	HOARSE	HORSE
GUN	GNU	KNEW
MARITAL	MARTIAL	MARSHAL
PLEAD	PEDAL	PEDDLE
TRIFLE	FILTER	PHILTRE
ASLEEP	PLEASE	PLEAS
RECALL	CELLAR	SELLER
LIGHTS	SLIGHT	SLEIGHT
WORDS	SWORD	SOARED
ARTIST	STRAIT	STRAIGHT
TISSUE	SUITES	SWEETS
SINEW	WINES	WHINES

(e)

#1	#2 = anagram of #1	#3 = constellation name of #2
RUGS	GRUS	CRANE
TRACER	CRATER	CUP
TROPIC	PICTOR	EASEL
SPICES	PISCES	FISH
CAMUS	MUSCA	FLY
PULES	LEPUS	HARE
MANOR	NORMA	LEVEL
CANTOS	OCTANS	OCTANT
RAISE	ARIES	RAM
LAVE	VELA	SAIL
HARDY	HYDRA	SERPENT
SCUNGY	CYGNUS	SWAN
NAMES	MENSA	TABLE
SCUTE	CETUS	WHALE

25.

SILENCE (Lambs) fits between COURT (Kangaroo) and EXALTATION (Lark).

Apes	Bear	Bee's	Black	Blue	Bull
PLANET	HUG	KNEES	GAME	THROAT	BUSH
Cat	Chick's	Cow	Crow	Dingo	Dog
FIDDLE	PEEP	PARSLEY	MURDER	DOING	DUCK
Donkey's	Elephant	Fox	Gnu	Goat	Green
YEARS	CASTLE	HOUNDS	GUN	COMPASSES	SHANK
Hare	Heron	Hornet	Horse's	Jackal	Jackal
LIP	SIEGE	THRONE	MOUTH	NIGHT	DAY
Kangaroo	**Lambs**	Lark	Leopard	Lion's	Mice	Monkey
COURT	**SILENCE**	EXALTATION	PAROLED	SHARE	NEST	SUIT
Nightingale	Owl	Phoenix	Pig	Rabbit	Ram's
WATCH	PARLIAMENT	FLIGHT	WHISTLE	PUNCH	HORN
Rat	Red	Shrew	Silver	Spring Hare	Viper's
ART	START	TAMING	BILL	REPHRASING	BUGLOSS
Vole	Whale	White	Wolf	Yellow	Zebra
LOVE	SCHOOL	CAP	HOUR	HAMMER	CROSSING

All themes relate to animals:

X of the Y: PLANET of the *APES*, NIGHT of the *JACKAL*, DAY of the *JACKAL*, **SILENCE** of the ***LAMBS***, FLIGHT of the *PHOENIX*, TAMING of the *SHREW*, HOUR of the *WOLF*

Can be preceded by a creature: *BEAR* HUG, *COW* PARSLEY, *HARE* LIP, *KANGAROO* COURT, *MONKEY* SUIT, *RABBIT* PUNCH, *ZEBRA* CROSSING

Can be preceded by the possessive of a creature: *BEE'S* KNEES, *CHICK'S* PEEP, *DONKEY'S* YEARS, *HORSE'S* MOUTH, *LION'S* SHARE, *RAM'S* HORN, *VIPER'S* BUGLOSS

Can be preceded by a colour to form an animal: *BLACK*GAME, *BLUE*THROAT, *GREEN*SHANK, *RED*START, *SILVER*BILL, *WHITE*CAP, *YELLOW*HAMMER

X and Y (pub name): *BULL* and BUSH, *CAT* and FIDDLE, *DOG* and DUCK, *ELEPHANT* and CASTLE, *FOX* and HOUNDS, *GOAT* and COMPASSES, *PIG* and WHISTLE

Collective noun: MURDER/*CROW*, SIEGE/*HERON*, EXALTATION/*LARK*, NEST/*MICE*, WATCH/*NIGHTINGALE*, PARLIAMENT/*OWL*, SCHOOL/*WHALE*

Anagram of a creature: DOING/*DINGO*, GUN/*GNU*, THRONE/*HORNET*, PAROLED/*LEOPARD*, ART/*RAT*, REPHRASING/*SPRING HARE*, LOVE/*VOLE*

26.

MY. (First, second, third and whole refer to the letters of MY.)

27.

(a) E-numbers: E464 + E123 − E412 = E175 = **Gold**

(b) Roman numerals removed from names of capital cities:

$$\frac{LIma + maniLa + moronI + sofIa - suCre}{VIenna - tunIs} \times roMe =$$

$$\frac{51 + 50 + 1 + 1 - 100}{6 - 1} \times 1000 = 600 = DC = \textbf{Washington}$$

(c) Dr Who actors:

$$\frac{5 + 10 \pm \sqrt{9}}{3} = 4 \text{ or } 6 = \textbf{Baker} \text{ (Tom or Colin)}$$

(d) Measurement conversions:

$$\frac{(14 \times 660) + 144 - 240}{3600} = 2.54 = \frac{\textbf{Inch}}{\textbf{cm}}$$

28.

(a) **10.** The phrase is an anagram of ONE, TWO, ..., TEN

(b) **GERMANY.** The phrase is an anagram of EINS, ZWEI, ..., ZEHN and (unlike, say, AUSTRIA) GERMANY has a sea border.

29.

CERTAINLY, PRIVATELY and **PUCKERING** are all 9-letter examples. We are aware of a few possible 10-letter examples: ANTICLERGY, CAPERINGLY, KERYGMATIC, LACQUEYING, PHLYCTENAR, TRIVALENCY and there are others.

30.

Encryption has been done by moving each letter along 13 places. Put another way, the following cipher alphabet has been used:

Plain: ABCDEFGHIJKLMNOPQRSTUVWXYZ
Cipher: NOPQRSTUVWXYZABCDEFGHIJKLM

The words become:

SENSE	PARTY	GUESS	HEAVEN	ESTATE	LADY	COLUMN
FRAFR	CNEGL	THRFF	URNIRA	RFGNGR	YNQL	PBYHZA

Which in order are:

(first) **LADY**, (second) **GUESS**, (third) **PARTY**, (fourth) **ESTATE**, (fifth) **COLUMN**, (sixth) **SENSE**, (seventh) **HEAVEN**.

31.

(97, 98). These are the next pair of consecutive numbers where the last letter of the first one is the same as the first letter of the last one.

32.

    (a) **143**. Cumulative totals of Fibonacci sequence

    (b) **6**. The sequence is the digits of pi, each plus one

33.

    (a) **Mae West**

    (b) **Bob Dole**

    (c) **Eric Bana**

    (d) **Bill Gates**

    (e) **Gus Van Sant**

    (f) **Ja Rule**

    (g) **Iman**

    (h) **Rod Laver**

    (i) **Meg Ryan**

    (j) **Ron Wood**

34.

QUESTION. TO BE OR NOT TO BE THAT IS THE is spelled out in the shape of a question mark. The word HAMLETS does not contribute to the shape, but is connected to the answer:

```
.    .   B   E   O   .   .
   T   O   .   R   N   .   .
.    .   .   .   .   O   T
.    .   .   T   O   .   .
.    .   .   B   E   .   .
.    .   T   H   .   .   .
.    .   .   A   T   .   .
.    .   I   S   .   .   .
.    .   .   .   .   .   .
   .   T   H   E   .   .   .
```

35.

(a)

#1	#2 = body part that precedes #1	#3 = anagram of #2
RING	EAR	ARE
WAVE	BRAIN	BAIRN
GREASE	ELBOW	BELOW
BURN	HEART	EARTH
MEAL	BONE	EBON
BAG	NOSE	EONS
PRINT	FINGER	FRINGE
PULL	LEG	GEL
STRAP	CHIN	INCH
FLINT	SKIN	INKS
CAP	KNEE	KEEN
VARNISH	NAIL	LAIN
CHAIR	ARM	MAR
SHOE	GUM	MUG
FLASK	HIP	PHI

(b)

#1	#2 = #1 less first letter	#3 = anagram of #2	#4 = #3 less first letter	#5 = anagram of #4	#6 = #5 less first letter
GAMBLE	AMBLE	BLAME	LAME	MALE	ALE
SPANS	PANS	SNAP	NAP	PAN	AN
GRASPING	RASPING	SPARING	PARING	RAPING	APING
LADDER	ADDER	DREAD	READ	DARE	ARE
STALE	TALE	LATE	ATE	EAT	AT
STABLE	TABLE	BLEAT	LEAT	LATE	ATE
CRATES	RATES	TEARS	EARS	SEAR	EAR
CHASTE	HASTE	HEATS	EATS	SEAT	EAT
RELAPSE	ELAPSE	ASLEEP	SLEEP	PEELS	EELS
SENDS	ENDS	SEND	END	DEN	EN
PRAISE	RAISE	ARISE	RISE	SIRE	IRE
BANGLED	ANGLED	DANGLE	ANGLE	GLEAN	LEAN
SNAILS	NAILS	SLAIN	LAIN	ANIL	NIL
ANODE	NODE	DONE	ONE	EON	ON
TROWEL	ROWEL	LOWER	OWER	WORE	ORE

36.

(a) **D**/ONE HUNDRE**D**: these are all the possible last letters of numbers in the order in which they first appear: ON**E**, TW**O**, FOU**R**, SI**X**, SEVE**N**, EIGH**T**, TWENT**Y**, ONE HUNDRE**D**.

(b) They are the lowest integers (UK-style) containing consecutive letters the same (thr**EE**, eigh**TT**housand, one mi**LL**ion, one millio**NN**inehundred, tw**OO**ctillion, onehundre**DD**ecillion); no numbers have this property with a different letter.

37.

(a) **VISITING**. Removing non-Roman leaves I, II, III, etc. V**Is**It**I**ng.

(b) **NAPE**. Parts of the body with first one or two letters forming the Periodic Table, N**A**pe.

(c) **UNQUESTIONED**. Middles spell out 'To be or not to be that is the question'.

(d) **ANSWER**. Middle letters give W, NW, N, NE, E, SE, S, an**SW**er.

(e) **OUST**. Words begin SUPER, CALI, FRAGIL, IS, TIC, EXPI, ALI, DOCI, OUS.

(f) **COMING**. Words contain DO, RE, MI rotating through. co**MI**ng.

38.

Mash-ups of plots of different films with the same title:

(a) **Rush** (1991 and 2013)

(b) **Carrie** (1952 and 1976/2013)

(c) **Rio** (1939 and 2011)

(d) **Fargo** (1952 and 1996)

(e) **Child's Play** (1972 and 1988)

(f) **The Host** (2006 and 2013)

(g) **Mr and Mrs Smith** (1941 and 2005)

(h) **Iron Man** (1951 and 2008)

(i) **The Trap** (1946 and 1966)

(j) **The Bodyguard** (1961 and 1992)

(k) **New Moon** (1930, 1940 and 2009)

(l) **Riff Raff** (1936, 1947, 1991 and 2009)

In (j), the 1961 Kurosawa film Yojimbo is also known as The Bodyguard, the English translation of its title.

39.

(a) **First 3 letters of US president's surname**: <u>AR</u>THUR, <u>BUS</u>H, <u>CAR</u>TER, <u>COO</u>LIDGE, <u>FOR</u>D, <u>HAY</u>ES, <u>KEN</u>NEDY, <u>MAD</u>ISON, <u>PIE</u>RCE, <u>WAS</u>HINGTON

(b) **Appear in country's capital**: PR<u>AIA</u>, Y<u>AOU</u>NDE, LJUBL<u>JA</u>NA, BA<u>GHD</u>AD, AS<u>HGA</u>BAT, BEI<u>JI</u>NG, SARA<u>JE</u>VO, NOUA<u>KCH</u>OTT, W<u>IND</u>HOEK, BA<u>NGK</u>OK, YAMOU<u>SSO</u>UKRO, BU<u>JU</u>MBURA, RE<u>YKJ</u>AVIK

(c) **Words in English and French**

(d) **Appear in a composer's surname**: RA<u>CHM</u>ANINOFF, RESP<u>IGH</u>I, MUSSOR<u>GSK</u>Y, BRA<u>HMS</u>, LI<u>SZ</u>T, SHOSTA<u>KOVIC</u>H, RIMS<u>KY-K</u>ORSAKOV, DE<u>LIUS</u>, MENDE<u>LSS</u>OHN, VA<u>UGH</u>AN-WILLIAMS

(e) **Appear within the word STEGANOGRAPHER**

(f) **If X is a word in the list then X-X is also a word**

(g) **International Vehicle Registration Codes**: AND/ANDORRA, BUR/MYANMAR, EAT/TANZANIA, ETH/ETHIOPIA, GUY/GUYANA, LAR/LIBYA, MAL/MALAYSIA, RIM/MAURITANIA, WAG/GAMBIA, WAN/NIGERIA

(h) **First 3 letters of a vegetable**: <u>ART</u>ICHOKE, <u>ASP</u>ARAGUS, <u>BEE</u>TROOT, <u>CAB</u>BAGE, <u>CAR</u>ROT, <u>LEE</u>K, <u>LET</u>TUCE, <u>MAR</u>ROW, <u>PAR</u>SNIP, <u>POT</u>ATO

(i) **Last 3 letters of a musical instrument**: DOUBLE B<u>ASS</u>, SACK<u>BUT</u>, ZI<u>THER</u>, ACCORD<u>ION</u>, CLARI<u>NET</u>, TROMB<u>ONE</u>, TRUM<u>PET</u>, D<u>RUM</u>, GUI<u>TAR</u>

(j) **First 3 letters of US states, reversed**: BEN/NEBRASKA, LAC/CALIFORNIA, LED/DELAWARE, NET/TENNESSEE, NEP/PENNSYLVANIA, NIM/MINNESOTA, RAM/MARYLAND, REV/VERMONT, SAW/WASHINGTON, SEW/WEST VIRGINIA

(k) **First, middle and last letters of countries**: <u>Alb</u>An<u>i</u>A, <u>Afgha</u>Nista<u>N</u>, <u>Ch</u>In<u>A</u>, <u>Ecu</u>Ado<u>R</u>, <u>Eri</u>Tre<u>A</u>, <u>In</u>Di<u>A</u>, <u>Leb</u>Ano<u>N</u>, <u>Les</u>Oth<u>O</u>, <u>Mor</u>Occ<u>O</u>, <u>Nam</u>Ibi<u>A</u>, <u>Sierr</u>A leon<u>E</u>, <u>Ukr</u>Ain<u>E</u>

(l) **Advancing (cyclically) each letter by 2 gives a word**: AME/*COG*, CEE/*EGG*, ELS/*GNU*, GAW/*ICY*, IGL/*KIN*, HYP/*JAR*, MUC/*OWE*, PYE/*RAG*, QIG/*SKI*, STY/*UVA*, UFW/*WHY*

(m) **ELM** ($5^2+12^2=13^2$): with A=1, B=2, ..., Z=26 the letters in each trigraph form Pythagorean triplets

(n) (before are **pre-nominal titles** with each letter replaced by its predecessor in the alphabet) **BOK**/CPL, **LQR**/MRS, **QDU**/REV, **RFS**/SGT, **RHQ**/SIR; (after are **post-nominal appellations** with each letter replaced by its successor in the alphabet) **FTR**/ESQ, **JJJ**/III, **KOS**/JNR, **PCF**/OBE, **TOS**/SNR

40.

THE LIST.

The words are anagrams of ROSE, LEEK, SHAMROCK and THISTLE, which are the symbols of the parts of the UK. THE LIST is an anagram of THISTLE.

41.

Rectangular.

When written columnwise in NATO phonetics, all the other words form squares. For example:

```
          A L P H A
          B R A V O                 G O L F              C H A R L I E
ABBOT =   B R A V O  ;   GLEE =     L I M A  ;   CUFF =  U N I F O R M
          O S C A R                 E C H O              F O X T R O T
          T A N G O                 E C H O              F O X T R O T
```

42.

   (a) **Sting**

   (b) **Halle Berry**

   (c) **Don Revie**

   (d) **Henry Fonda**

   (e) **Man Ray**

   (f) **Mia Hamm**

   (g) **Teri Hatcher**

   (h) **Tone Loc**

   (i) **Ang Lee**

   (j) **J Lo**

Variant answers: (g) **Al Read**, (i) **Una Lee** or **Eve Arden**, (j) **C No**.

43.

   (a) **Anagrams**: AGREES/GREASE, ASPIRE/PRAISE, CHASTE/ SACHET, ENLIST/SILENT, ENTRAP/PARENT, LATTER/ RATTLE, NEURAL/UNREAL, PALEST/STAPLE, PRIEST/ STRIPE,

   (b) **Concatenate to form a new word**: ASP/HALT, BAR/GAIN, CHOPS/TICK, COME/DIES, CUP/BOARD, DAM/NATION, MILES/TONE, REST/RAIN, SIDE/REAL

   (c) **Rhymes**: ALTHOUGH/GATEAU, BEER/TIER, BITE/ SLEIGHT, BLUE/THROUGH, CONFINE/RESIGN, DISDAIN/SANE, LAUGHED/RAFT, LYNX/WINKS, QUARTZ/WARTS

   (d) **Nest to form a new word**: BA/THE/R, BO/WIN/G, FO/RAG/E, LE/GAT/E, PO/TEN/T, RE/FUN/D, RO/BUS/T, SH/ANT/Y, TR/END/Y, UR/BAN/E

   (e) **Interleave to form a new word**: AIL/FED/*AFIELD*, AIR/PAY/ *APIARY*, ALE/LID/*ALLIED*, ASS/SIT/*ASSIST*, BUS/ORE/ *BOURSE*, EFT/FEE/*EFFETE*, FIN/RED/*FRIEND*, FIR/RAY/ *FRIARY*, FUN/LET/*FLUENT*, JUT/ANY/*JAUNTY*, LUG/ONE/ *LOUNGE*, OIL/ROE/*ORIOLE*, PIE/AND/*PAINED*, SET/WAY/ *SWEATY*, SOT/PRY/*SPORTY*, TIE/OLD/*TOILED*, TOP/RUE/ *TROUPE*, WOE/ODD/*WOODED*

(f) **Anagrams of European countries**: AIL+VAT=LATVIA, ANY+ROW=NORWAY, CAR+FEN=FRANCE, CUR+SPY=CYPRUS, END+SEW=SWEDEN, KEY+RUT=TURKEY, OLD+PAN=POLAND

(g) **Anagrams of US states**: ADORN+SHIELD=RHODE ISLAND, AID+OH=IDAHO, AIM+SONNET=MINNESOTA, AMOK+ HALO=OKLAHOMA, ANT+MOAN=MONTANA, AT+SEX=TEXAS, BEAR+SANK=NEBRASKA, COOL+ ROAD=COLORADO, FACIAL+IRON=CALIFORNIA, FOR+LAID=FLORIDA, GORE+ON=OREGON, HATING+ SNOW=WASHINGTON, JEWRY+SEEN=NEW JERSEY, LARD+MANY=MARYLAND, REAL+WADE= DELAWARE, SERAPHIM+WHEN=NEW HAMPSHIRE, STAIR+ VIEWING=WEST VIRGINIA, WORK+YEN=NEW YORK

44.

Reading across the matrices one letter at a time starting top left then middle row left, etc. gives: JUST COMPLETE THE FINAL MATRIX WITH THE MISSING LETTERS FROM THIS SENTENCE.

$$\begin{pmatrix} M & I & R \\ H & I & S \\ M & E & E \end{pmatrix}$$

45.

One-word film titles and their casts:

Alien: Sigourney Weaver, Tom Skerritt, Harry Dean Stanton, John Hurt, Yaphet Kotto

Brazil: Jonathan Pryce, Robert de Niro, Michael Palin, Bob Hoskins, Kim Greist

Cocoon: Don Ameche, Jessica Tandy, Hume Cronyn, Tahnee Welch, Steve Guttenberg

Deliverance: Jon Voight, Burt Reynolds, Ned Beatty, Ronny Cox

Excalibur: Helen Mirren, Nigel Terry, Cherie Lunghi, Nicholas Clay

Flight: Denzel Washington, Don Cheadle, Kelly Reilly, John Goodman, Melissa Leo

Gravity: Sandra Bullock, George Clooney, Ed Harris, Orto Ignatiussen, Paul Sharma, Amy Warren

Halloween: Jamie Lee Curtis, Donald Pleasence, P. J. Soles, Nancy Loomis

Inception: Leonardo DiCaprio, Ken Watanabe, Marion Cotillard, Ellen Page, Tom Berenger

Jaws: Roy Scheider, Robert Shaw, Richard Dreyfuss, Lorraine Gary, Murray Hamilton

Krull: Lysette Anthony, Alun Armstrong, Francesca Annis, Liam Neeson, Robbie Coltrane

Labyrinth: Jennifer Connelly, David Bowie, Toby Froud, Christopher Malcolm, Shelley Thompson

Madagascar: Ben Stiller, Chris Rock, David Schwimmer, Jada Pinkett Smith, Sacha Baron Cohen

Nixon: Anthony Hopkins, Joan Allen, Paul Sorvino, Bob Hoskins, E. G. Marshall

Octopussy: Roger Moore, Maud Adams, Steven Berkoff, Desmond Llewelyn, Kristina Waybourn

Philadelphia: Tom Hanks, Denzel Washington, Jason Robards, Antonio Banderas, Joanne Woodward

Quadrophenia: Phil Daniels, Ray Winstone, Leslie Ash, Toyah Wilcox, Sting

Rollerball: James Caan, John Houseman, Maud Adams, John Beck, Ralph Richardson

Skyfall: Daniel Craig, Javier Bardem, Ralph Fiennes, Berenice Marlohe, Judi Dench

Trainspotting: Ewan McGregor, Ewen Bremner, Johnny Lee Miller, Kevin McKidd, Robert Carlyle

Unforgiven: Clint Eastwood, Gene Hackman, Morgan Freeman, Richard Harris

Venus: Peter O'Toole, Leslie Phillips, Jodie Whitaker, Richard Griffiths, Vanessa Redgrave

Westworld: Yul Brynner, Richard Benjamin, James Brolin

Xanadu: Olivia Newton-John, Michael Beck, Gene Kelly, James Sloyan, Dimitra Arliss

Yellowbeard: Graham Chapman, Peter Cook, Eric Idle, John Cleese, Spike Milligan

Zoolander: Ben Stiller, Owen Wilson, Will Ferrell, Milla Jovovich, Jon Voight

There are some alternative cast members with matching initials (e.g. Michael Caine instead of Marion Cotillard for Inception).

46.

EACH OF THE NUMBERS BETWEEN TWO AND NINE ON A MOBILE PHONE CORRESPONDS TO THREE OR FOUR LETTERS OF THE ALPHABET AND EACH LETTER IN THIS TEXT IS RANDOMLY REPLACED BY ANOTHER WHICH CORRESPONDS TO THE SAME NUMBER AS THAT LETTER

47.

Tip: Try typewriter, top row.

The letters have been enciphered by using the numbers which appear above them on a typewriter.

```
TIP: TRY TYPEWRITER, TOP ROW

580: 546 5603248534, 590 492
```

48.

HEAVEN (Seventh) fits between HAND (Second) and DARK (Shot in the).

Afghan	Black	Blue	Brown	Chin
HOUND	ART	BLOOD	BREAD	STRAP

Chinese	Cuban	Danish	Dog	Drop
LANTERN	HEEL	PASTRY	MANGER	BUCKET

Dyed	Ear	Elbow	Fifth	First
WOOL	WIG	GREASE	COLUMN	CLASS

Fly	Foot	Fourth	French	Green
OINTMENT	SOLDIER	DIMENSION	LEAVE	BELT

Grey	Hand		Hip	Indian	Jack
MATTER	SHAKE		FLASK	SUMMER	BOX
Knee	Maltese		Man	Nineteenth	Norwegian
CAP	CROSS		MOON	HOLE	WOOD
Nose	Orange		Pain	Pie	Pink
BAG	PEEL		NECK	SKY	GIN
Red	Second	**Seventh**	Shot	Shoulder	Sixth
BRICK	HAND	**HEAVEN**	DARK	BLADE	SENSE
Spanish	Swiss		Third	Toad	Twelfth
FLY	ROLL		WORLD	HOLE	NIGHT
Twentieth	White		Wrist	Yellow	
CENTURY	COLLAR		WATCH	PAGES	

Themes are:

Country adjectives: Afghan HOUND, Chinese LANTERN, Cuban HEEL, Danish PASTRY, French LEAVE, Indian SUMMER, Maltese CROSS, Norwegian WOOD, Spanish FLY, Swiss ROLL

Colours: Black ART, Blue BLOOD, Brown BREAD, Green BELT, Grey MATTER, Orange PEEL, Pink GIN, Red BRICK, White COLLAR, Yellow PAGES

Parts of the body: Chin STRAP, Ear WIG, Elbow GREASE, Foot SOLDIER, Hand SHAKE, Hip FLASK, Knee CAP, Nose BAG, Shoulder BLADE, Wrist WATCH

X in the Y: Dog (in the) MANGER, Drop (in the) BUCKET, Dyed (in the) WOOL, Fly (in the) OINTMENT, Jack (in the) BOX, Man (in the) MOON, Pain (in the) NECK, Pie (in the) SKY, Shot (in the) DARK, Toad (in the) HOLE

Ordinals: Fifth COLUMN, First CLASS, Fourth DIMENSION, Nineteenth HOLE, Second HAND, **Seventh HEAVEN,** Sixth SENSE, Third WORLD, Twelfth NIGHT, Twentieth CENTURY

49.

    (a) **ABBA**

Waterloo

Ring Ring

Money, Money, Money

Knowing Me, Knowing You

The Name Of The Game

Does Your Mother Know

Summer Night City

Super Trouper

Chiquitita

    (b) **BEATLES**

Yesterday

Hey Jude

Let It Be

Can't Buy Me Love

All You Need Is Love

A Hard Day's Night

Strawberry Fields Forever

Get Back

Help!

    (c) **BEACH BOYS**

Cottonfields

Good Vibrations

God Only Knows

Wouldn't It Be Nice

Heroes And Villains

Barbara Ann

Darlin'

(d) **ROLLING STONES**

Angie

Brown Sugar

The Last Time

It's All Over Now

Get Off Of My Cloud

Undercover Of The Night

Paint It, Black

Tumbling Dice

Respectable

(e) **QUEEN**

Flash

Killer Queen

Somebody To Love

We Are The Champions

Crazy Little Thing Called Love

Seven Seas Of Rhye

Radio Ga Ga

One Vision

Breakthru'

50.

451AD: *Fahrenheit 451* is by Ray *Bradbury*. Malcolm *Bradbury*, from Sheffield, set his novel *The History Man* in a fictional town '*Watermouth*'. The real *Watermouth Castle* is really a country house, designed by *George Wightwick*, who worked in Plymouth with *John Foulston*, who died on 30 December 1841. On that day, *Ralph Waldo Emerson* gave his first lecture on manners, the 7th being given on 13 January 1842, the same day that *William Brydon* returned as the sole survivor of the retreat from Kabul. His retreat was painted by *Lady Elizabeth Butler*, whose husband *Sir William Butler* wrote the biography of *George Colley*, who was killed at *Majuba Hill*, the last battle of the *First Boer War*. The peace treaty was signed on 23 March 1881, the birth date of *Roger Martin du Gard* (Literature Nobel Laureate) and *Hermann Standinger* (Chemistry Nobel Laureate), who was born in *Worms*. The Diet of Worms was presided over by *Charles V*, who was crowned King in *Aachen Cathedral*, which was built by *Charlemagne*, who was authorised by *Pope Adrian I* to take what he wanted from *Ravenna*, which is where *Valentinian III* was born whose daughter *Eudocia* married *Huneric*, King of the Vandals, who was previously married to the daughter of *Theodoric I*, who died at the Battle of Châlons in **451AD**. Thus you have 451 at both ends of the chain!

51.

(a) Chemical formulae: $C_2H_4O_2 + C_6H_6O + CO_2 - C_9H_8O_4 = H_2O =$ **Water**

(b) Champagne bottle sizes: $(20 \times 8) / (24 + 16) = 4 =$ **Jeroboam**

(c) Sports team sizes: $(9 \times 11 \times 7) - (5 \times 15 \times (13 - 4)) = 18 =$ **Australian Rules Football**

52.

(a) **107.** The numbers are increased by the sum of their digits. $94 + 13 = 107$

(b) **102.** The numbers are increased by the product of their digits. $102 + 0 = 102$

(c) **44.** The numbers are increased by the difference of their digits. All the numbers in the rest of the sequence are 44.

53.

(a) SOYLENT **GREEN** CARD

(b) DIE **HARD TIMES** SQUARE

(c) UNIVERSAL **SOLDIER BLUE STEEL** MAGNOLIAS

(d) CABIN **FEVER PITCH BLACK RAIN** MAN

(e) NEAR **DARK ANGEL HEART BEAT GIRL** INTERRUPTED

54.

CNA. Reversing the list of trigraphs and reading the middle letters gives:

WHICH THREE LETTER GROUP PRECEDES CZN IN THIS QUESTION? The answer to this question is CNA.

In the question code, the first letter of each trigraph is C if the plain letter is a consonant and V if it's a vowel. The last letter of the trigraph is derived from:

Consonant: BCDFGHJKLMNPQRSTVWXYZ   Vowel: AEIOU
Last letter: ABCDEFGHIJKLMNOPQRSTU   Last letter: ABCDE

Decoding the question gives: YOU DO NOT NEED TO BREAK THE CODE TO ANSWER THIS QUESTION – which is true!

55.

These are the last letters of the names of the letters of the Greek alphabet (alph**a**, bet**a**, gamm**a**, delt**a**, epsilo**n**, etc.).

56.

(a) **NORTH DAKOTA:** The list of US states contain the letters A, B, etc. up to Z; then A, B, etc. up to H. All the other remaining states (CONNECTICUT, HAWAII, ILLINOIS, IOWA, MAINE, MINNESOTA, MISSOURI, NEW HAMPSHIRE, NEW MEXICO, OHIO, PENNSYLVANIA, RHODE ISLAND, SOUTH CAROLINA, WASHINGTON, WISCONSIN, WYOMING) contain the letter I.

(b) **CENTRAL AFRICAN REPUBLIC:** It's the only remaining landlocked country in Africa in a list that alternates non-landlocked/landlocked.

57.

(a) **Who Framed Roger Rabbit**

(b) **Whatever Happened to the Likely Lads?**

(c) **What Difference Does It Make?** – The Smiths

(d) **Why Didn't They Ask Evans?** – Agatha Christie

(e) **Dude, Where's My Car?**

(f) **What Ever Happened to Baby Jane?**

(g) **What Shall We Do with the Drunken Sailor?**

(h) **What Does the Fox Say?** – Ylvis

(i) **Where in the World Is Carmen Sandiego?** – videogame

58.

MONDAY.

The missing words are VEERED, TENDER, STAYED, REMOVE, RETEST and the six words form a 3×3 box:

RE TE ST

MO ND AY

VE ER ED

59.

They are military ranks minus their relevant abbreviation:

brigADIER captAIN genERAL sERgEANt admIRAL cOMMANdEr cOMMOdOre colONEL majOR corpORAL pRIVAte

60.

DROP (h's), **DOT** (i's), **MIND** (p's and q's), **ROLL** (r's), **CROSS** (t's)

61.

In English, there aren't any pairs. The property is that of consecutive pairs of integers having no letters in common. In French: (DEUX, TROIS), (NEUF, DIX), (DIX, ONZE). In German: (VIER, FÜNF), (FÜNF, SECHS), (SIEBEN, ACHT), (ACHT, NEUN).

62.

COSY (Tea) fits between LAZY (Susan) and BIRDS (Thunder).

Acres	Archimedes'	Asking		Atlanta	Baltimore
SCARE	PRINCIPLE	YOURS		CONSTITUTION	SUN
Beer	Beethoven's	Best		Boston	Brandy
GARDEN	FIFTH	HOPE		GLOBE	SNAP
Buck	Bushels	Chain		Champagne	Cloud
BANG	BLUSHES	CHINA		PRESS	BURST
Coffee	Columbus	Course		Fog	Future
BEAN	DISPATCH	PAR		HORN	PLAN
Hail	Hectare	Hobson's		Hurricane	Inch
STONE	TEACHER	CHOICE		LAMP	CHIN
Jenny	Jugular	Maria		Mark	Memories
SPINNING	GO	BLACK		QUESTION	THANKS
Mercator's	Miami	Miles		Milk	Montezuma's
PROJECTION	HERALD	SLIME		TOOTH	REVENGE
Pandora's	Pascal's	Peter		Pole	Rain
BOX	TRIANGLE	BLUE		LOPE	CHECK
Road	Robin	Rose		Rubik's	Rum
ONE	ROUND	DOG		CUBE	RUNNER
Sacramento	Snow	Susan	**Tea**	Thunder	USA
BEE	PLOUGH	LAZY	**COSY**	BIRDS	TODAY
Washington	Water	William		Wind	Yards
POST	FOUNTAIN	SWEET		SOCK	DRAYS

Themes are:

Anagrams of units of measurements: SCARE/*ACRES*, BLUSHES/*BUSHELS*, CHINA/*CHAIN*, TEACHER/*HECTARE*, CHIN/*INCH*, SLIME/*MILES*, LOPE/*POLE*, DRAYS/*YARDS*

Can be preceded by the possessive of a name: *ARCHIMEDES'* PRINCIPLE, *BEETHOVEN'S* FIFTH, *HOBSON'S* CHOICE,

MERCATOR'S PROJECTION, *MONTEZUMA'S* REVENGE, *PANDORA'S* BOX, *PASCAL'S* TRIANGLE, *RUBIK'S* CUBE

X for the Y: YOURS for the *ASKING*, HOPE for the *BEST*, BANG for the *BUCK*, PAR for the *COURSE*, PLAN for the *FUTURE*, GO for the *JUGULAR*, THANKS for the *MEMORIES*, ONE for the *ROAD*

Can be preceded by a US city to form the name of a US newspaper: *ATLANTA* CONSTITUTION, *BALTIMORE* SUN, *BOSTON* GLOBE, *COLUMBUS* DISPATCH, *MIAMI* HERALD, *SACRAMENTO* BEE, *USA* TODAY, *WASHINGTON* POST

Can be preceded by a drink: *BEER* GARDEN, *BRANDY* SNAP, *CHAMPAGNE* PRESS, *COFFEE* BEAN, *MILK* TOOTH, *RUM* RUNNER, **_TEA_ COSY**, *WATER* FOUNTAIN

Can be followed by a Christian name: SPINNING *JENNY*, BLACK *MARIA*, QUESTION *MARK*, BLUE *PETER*, ROUND *ROBIN*, DOG *ROSE*, LAZY *SUSAN*, SWEET *WILLIAM*

Can be preceded by a weather phenomenon: *CLOUD*BURST, *FOG*HORN, *HAIL*STONE, *HURRICANE* LAMP, *RAIN*CHECK, *SNOW* PLOUGH, *THUNDER*BIRDS, *WIND*SOCK

63.

(a) Chinese Years: $9 + 5 + 4 - (11 + 1) = 6 =$ **Snake**. We've started with Rat $= 1$, but the answer is independent of how you start the numbering.

(b) Words consisting of two letters followed by a number: (OFten $\times$ EAten $-$ FReight $\times$ SLeight) / (CRone + ALone + SHone + OZone) $= (10 \times 10 - 8 \times 8) / (1 + 1 + 1 + 1) = 9 =$ CAnine $=$ **CA**

(c) The Lord of the Rings: $(4+3+7+2) \times 9 = 144 =$ **Guests at birthday party** (or **Combined ages of Bilbo and Frodo**)

(d) Words with numbers removed from the middle (leaving two letters before and two after): ((AnTENna + PuNINEss $-$ LiONEss) / NeTWOrk) $-$ PiONEer $= ((10 + 9 - 1) / 2) - 1 = 8 =$ FrEIGHTer $=$ **Frer** (or FrEIGHTed $=$ **Fred**)

64.

BABY (for example) fits between OIL and BALL. All the words in the list can be preceded by words beginning C-: in particular CA-, CH-, CO-, CR- and CU-; for each of these digraphs there are seven words of lengths 3–9 letters, one word of each length.

Calamity	Camp	Capital	Car	Carbon
JANE	FOLLOWER	CITY	PET	DATING

Carol	Catherine	Chambers	Che	Check
SINGER	WHEEL	DICTIONARY	GUEVARA	MATE

Cheese	Chicken	Chop	Christmas	Coat
GRATER	RUN	SUEY	CAROL	HANGER

Cocker	Collected	Corporal	Cottage	Couch
SPANIEL	WORKS	PUNISHMENT	PIE	POTATO

Cow	Cranberry	Credit	Criminal	Crop
BOY	SAUCE	CARD	RECORD	CIRCLE

Crude	**Cry**	Crystal	Cube	Cucumber	Cup
OIL	**BABY**	BALL	ROOT	FRAME	BOARD

Curiosity		Current	Cuttle	Cutty
SHOP		AFFAIRS	FISH	SARK

CA: *CALAMITY* JANE, *CAMP* FOLLOWER, *CAPITAL* CITY, *CAR*PET, *CARBON* DATING, *CAROL* SINGER, *CATHERINE* WHEEL

CH: *CHAMBERS* DICTIONARY, *CHE* GUEVARA, *CHECK*MATE, *CHEESE* GRATER, *CHICKEN* RUN, *CHOP* SUEY, *CHRISTMAS* CAROL

CO: *COAT* HANGER, *COCKER* SPANIEL, *COLLECTED* WORKS, *CORPORAL* PUNISHMENT, *COTTAGE* PIE, *COUCH* POTATO, *COW*BOY

CR: *CRANBERRY* SAUCE, *CREDIT* CARD, *CRIMINAL* RECORD, *CROP* CIRCLE, *CRUDE* OIL, *CRY* BABY, *CRYSTAL* BALL

CU: *CUBE* ROOT, *CUCUMBER* FRAME, *CUP*BOARD, *CURIOSITY* SHOP, *CURRENT* AFFAIRS, *CUTTLE*FISH, *CUTTY* SARK

65.

(a) The numbers are the minimum number of moves a knight (starting in the normal starting square in the bottom left) would take to reach that square:

2 3 2 3

1 2 1 4

2 3 2 1

3 0 3 2

(b) The numbers are the minimum number of moves a pawn (starting in a normal starting square for a rook's pawn in the bottom left) would take to reach that square:

1 2 2 6

1 1 6 6

0 6 6 6

6 6 6 6

66.

(a) Verve	paired with	**(l) The Verve**
(b) The	paired with	**(f) The The**
(c) Hague	paired with	**(h) The Hague**
(i) Misfits	paired with	**(d) The Misfits**
(j) 'Stand'	paired with	**(e) The Stand**
(k) Sting	paired with	**(g) The Sting**

The band 'The Verve' was called 'Verve' up until 1994.

67.

14. The number contains the letter at the corresponding position in the alphabet:

FIVE, EIGHT, NINE, TWELVE, FOURTEEN, TWENTY, TWENTY-THREE, TWENTY-FIVE

68.

The 3 remaining words are **BANK, FIELD** and **OVERS,** all of which can be preceded by **LEFT.**

The pairings are:

Vegetables: POT, A, TO and MAN, GET, OUT

Kings and Queens: ARTHUR, ELVIS, KONG and ANNE, DANCING, VICTORIA

Anagrams: ASCERTAIN, CARTESIAN, SECTARIAN and INTEGRAL, RELATING, TRIANGLE

Dutch/Spanish: AUCTION, BARN, COURAGE and FLY, MAIN, STEPS

Door/Window: BELL, MAT, STEP and CLEANER, SHOPPING, SILL

Battles/Treaties: BLENHEIM, BRITAIN, SLUYS and ROME, VERSAILLES, VIENNA

Sherlock Holmes adventures: HIS, LAST, BOW and GABLE, GARRIDEB, STUDENT

Days/Nights: BOXING, GROUNDHOG, MAY and BURNS, SILENT, TWELFTH

Cambridge/Oxford colleges: CHURCHILL, DOWNING, DARWIN and LINCOLN, ORIEL, UNIVERSITY

Monopoly/Cluedo pieces: DOG, HAT, IRON and MUSTARD, PEACOCK, PLUM

69.

(a) **ROSE** (hip) **BATH**; **BLACK** (eye) **BALLS**; **BOOT** (leg) **IRON**; **YELLOW** (belly) **BUTTON**; **RIGHT** (arm) **CHAIR**; **COLD** (shoulder) **PADS**; **TENNIS** (elbow) **GREASE**; **TEA** (chest) **NUT**; **RING** (finger) **TIPS**; **SECOND** (hand) **SPRING**

(b) **BLOOD** (orange) **PEEL**; **NAVY** (blue) **BOTTLE**; **SEEING** (red) **BREAST**; **EARL** (grey) **HOUND**; **RACING** (pink) **ELEPHANTS**; **LINCOLN** (green) **HOUSE**; **GERMAN** (silver) **LINING**; **JET** (black) **MARKET**; **SNOW** (white) **WASH**

70.

The answers are: BAKING POWDER, FLYING SAUCER, JACK-IN-THE-BOX, MOTHERS-IN-LAW, NOTARY PUBLIC, PANTY GIRDLES, PLASTIC MONEY, PLAYS FOR TIME, PLOUGHED BACK, QUESTION MARK, SHADOW FIGURE, STAYING POWER.

(a) **All answers contain 12 letters and all the letters are different**.

(b) **TALCUM POWDER** (for example).

71.

(a) ALLEY, DESERT, POLAR, CRY, HOT, LOAN, MARCH, SEA, SCREECH, SHIRE, ROLLER

Precede CAT, RAT, BEAR, WOLF, DOG, SHARK, HARE, LION, OWL, HORSE, SKATE

Which anagram to:

ACT, ART, BARE, FLOW, GOD, HARKS, HEAR, LINO, LOW, SHORE, TAKES

(b) THE HAGUE, VALETTA, ANKARA, MADRID, HAVANA, PARIS, LISBON, COPENHAGEN, BERN, MOSCOW, NEW DELHI, BEIJING

are capitals of countries with adjectives: DUTCH, MALTESE, TURKISH, SPANISH, CUBAN, FRENCH, PORTUGUESE, DANISH, SWISS, RUSSIAN, INDIAN, CHINESE which precede: COURAGE, CROSS, DELIGHT, FLY, HEEL, LEAVE, MAN-OF-WAR, PASTRY, ROLL, ROULETTE, SUMMER, WHISPERS

(c) CENT, HEAVEN, FOOL, HAMLET, ATTITUDE, MECHANICAL, OUTRAGEOUS, DIOCESE, VOLITION, BRASSICA, RIPE

are synonyms for: PENNY, SKY, CHARLIE, VILLAGE, MOOD, CLOCKWORK, SHOCKING, SEE, WILL, CABBAGE, MELLOW

which precede: BLACK, BLUE, BROWN, GREEN, INDIGO, ORANGE, PINK, RED, SCARLET, WHITE, YELLOW

(d) RICH, HAIRY, WARM, CROOKED, ALIVE, DARK, UNWELL, SANE, LOW, UGLY, FAT

are antonyms of: POOR, BALD, COOL, STRAIGHT, DEAD, LIGHT, FIT, MAD, HIGH, PRETTY, THIN

which can be compared to (X as a Y): CHURCHMOUSE, COOT, CUCUMBER, DIE, DODO, FEATHER, FIDDLE, HATTER, KITE, PICTURE, RAKE

(e) MOCK, BED, WAG, TELL, WRAITH, FROM, TOP, METRE, NAME, RUNG, BOOK

rhyme with: LOCK, RED, RAG, BELL, FAITH, TOM, HOP, PETER, GAME, HUNG, HOOK

which are the first parts of threes ending: BARREL, BLUE, BOBTAIL, CANDLE, CHARITY, HARRY, JUMP, MARY, MATCH, QUARTERED, SINKER

72.

(a) **PACK**. Preceding numbers, or the word 'point', give 3.1415926.

(b) **SHRINKING**. Associated colours form the rainbow. Shrinking Violet.

(c) **CART**. Words can be preceded/followed by animals in alphabetical order. *ANT*-ELOPE, BUG-*BEAR*, *CARP*-ENTER, REIN-*DEER*, *EMU*-LATE, OUT-*FOX*, *GOAT*-SUCKER, CART-*HORSE*.

73.

    (a) **Nat King Cole**

    (b) **Paul Simon**

    (c) **Pele**

    (d) **Macy Gray**

    (e) **Eva Peron**

    (f) **Martin Shaw**

    (g) **Cary Grant**

    (h) **Sally Ride**

    (i) **Tom Hanks**

    (j) **Ed Wood**

74.

ME/I/1, AGAINST/V/5, VIOLET/VI/6, KISS/X/10, GREEK LETTER/XI/11, 587 YARDS (approx)/LI/51, CARBON/C/100, LIFE STORY/CV/105, SRI LANKA/CL/150, COMPACT DISC/CD/400, RALPH FIENNES/M/1000

75.

    (a) 'I Saw Mommy Kissing Santa Claus'

    (b) 'In the Bleak Midwinter'

    (c) 'Merry Christmas, Darling'

    (d) 'Silent Night'

    (e) 'Winter Wonderland'

    (f) 'Little Drummer Boy'

    (g) 'Angels We Have Heard on High'

    (h) '2000 Miles'

    (i) 'We Three Kings of Orient Are'

    (j) 'Let It Snow'

76.

The properties of the words to the left of the colon are:

(a) **Consists of 2 letters adjacent in the alphabet, and 2 letters 1 apart**.

(b) **None of the capital cities is the largest in its country**.

(c) **Each contains, as consecutive letters, the anagram of a letter in the NATO phonetic alphabet**.

(d) **Each can be divided into 3 words (e.g. AM-BIT-ION)**.

(e) **Each can be pronounced in 2 different ways**.

(f) **Each name contains another name written backwards within**.

(g) **Each word remains a word if R ↔ L, L ↔ R**.

77.

The archivist has added the letters of the names (A=1, B=2, etc.) to each other when there is an AND in the title. So for example U2's RATTLE and HUM yields R, A, T, B(=T+H), G(=L+U), R(=E+M)

(a) **BUTCH CASSIDY AND THE SUNDANCE KID**

(b) **X&Y**

(c) **PRIDE AND PREJUDICE AND ZOMBIES**

78.

P(APA). M(IKE) lives in Q(UEBEC), and is very fond of G(OLF). He is married to J(ULIET), and they have two children, C(HARLIE) and V(ICTOR). What do the children call M(IKE)? These are all letters in the NATO phonetic alphabet.

79.

(a) **CONTEMPORANEOUS** has a Scrabble score of 21, the others 17.

(b) All 98 letter Scrabble tiles are used, plus an extra C and T, so that if blanks are used for these (thus using all 100 tiles) **the score of CONTEMPORANEOUS is also 17**.

(c) **17** (the total score of all the tiles – which must be divisible by the number of words – is 187 = 11 × 17) An example would be: BABY, CONTINUES, FEWER, FULLY, GUARDING, HEAVE, IRRIGATION, JOIN, MUSICAL, OXEN, QUIT (blanks for UI), RETALIATED, ROMPED, SOOTHED, SPEAK, WAIVE, ZO.

80.

(Whati Stent Imese Leven) – (Pound Sina Stone)

= What is ten times eleven – pounds in a stone

= $10 \times 11 - 14 = 96$ = Ninety Six = **34°10'N 82°01'W**

81.

Gobbledygook. From the second line onwards letters in the top row of a typewriter keyboard have moved one place to the left. From the third line the same has happened to letters in the second row of the keyboard, and in the last line it has happened to all letters.

```
SOMETHING PECULIAR HAPPENED WHEN I SAT DOWN

TO USE THE TYPEWRITER. THE FAULT GOT WORSE
RI YSW RHW RTOWQEURWE. RHW FAYLR GIR QIESW

UNTIL THE TEXT COULD NOT BE RECOGNISED. IT
YNRUK RGW RWXR CIYKS NIR BW EWCIFNUAWS. UR

SEEMED TO BE COMPLETE GOBBLEDYGOOK.
AWWNWS RI VW XINOKWRW FIVVKWSTFIIJ.
```

82.

Countries externally bordering the given country, going clockwise:

Afghanistan	=	Turkmenistan, Uzbekistan, Tajikistan, China, Pakistan, Iran
Bangladesh	=	India, Myanmar
Chile	=	Peru, Bolivia, Argentina
Djibouti	=	Ethiopia, Eritrea, Somalia
Ecuador	=	Colombia, Peru
France	=	Belgium, Luxembourg, Germany, Switzerland, Italy, Monaco, Spain, Andorra, Spain
Greece	=	Albania, Macedonia, Bulgaria, Turkey
Hungary	=	Austria, Slovakia, Ukraine, Romania, Serbia, Croatia, Slovenia
Italy	=	France, Switzerland, Austria, Slovenia

Japan/Jamaica	=	[no land borders with other countries]
Kyrgyzstan	=	Kazakhstan, China, Tajikistan, Uzbekistan
Liechtenstein	=	Austria, Switzerland
Moldova	=	Ukraine, Romania
Nigeria	=	Niger, Chad, Cameroon, Benin
Oman	=	Yemen, Saudi Arabia, **United Arab Emirates**
Portugal	=	Spain
Qatar	=	Saudi Arabia
Rwanda	=	Uganda, Tanzania, Burundi, Democratic Republic of the Congo
Slovenia	=	Austria, Hungary, Croatia, Italy
Tunisia	=	Libya, Algeria
Uruguay	=	Brazil, Argentina
Venezuela	=	Guyana, Brazil, Colombia
W	=	[no single-word country starting with a W]
X	=	[no country starting with an X]
Yemen	=	Saudi Arabia, Oman
Zimbabwe	=	Zambia, Mozambique, South Africa, Botswana

83.

(a) **DOG**: all others can be formed from the letters of CHRISTMAS.

(b) **FEL**: in all others the central vowel can be any vowel and a word is still formed.

84.

(a) <u>Crocodile</u> Dundee

(b) <u>Tiger</u> Woods

(c) <u>Rooster</u> Cogburn

(d) <u>Cat</u> Stevens

(e) <u>Robin</u> Hood

(f) <u>Newt</u> Gingrich

85.

    (a) First letters of Pi (Three Point One Four One Five Nine Two Six Five Three, etc.): **T**

    (b) For n=3, 4, 5, ..., the smallest number that is divisible by n and has n letters: **300, -, 70000000, 15000, 14000** (no solution for n=13)

86.

MILLENNIUM. The other answers are SECOND, MINUTE, HOUR, DAY, WEEK, MONTH, YEAR, DECADE, CENTURY.

87.

    (i)

        (a) **OGDEN NASH**: (the phrase is 'CANDY IS DANDY BUT LIQUOR IS QUICKER', enciphered by simple substitution using the name as key-phrase

        (b) **MARK TWAIN**: (similarly, 'GOLF IS A GOOD WALK SPOILED')

        (c) **OSCAR WILDE**: ('I CAN RESIST EVERYTHING EXCEPT TEMPTATION')

        (d) **BENJAMIN DISRAELI**: ('LIES, DAMNED LIES AND STATISTICS')

        (e) **EDWARD BULWER-LYTTON**: ('THE PEN IS MIGHTIER THAN THE SWORD')

        (f) **THEODORE ROOSEVELT**: ('SPEAK SOFTLY AND CARRY A BIG STICK')

    (ii) **YJU RAG LJJE** ('YOU CAN FOOL ...', a quote attributed to ABRAHAM LINCOLN)

88.

IN. Add up adjacent letters (mod 26) and subtract 1. E.g. for RIGHTS → ZONAL:

R(18)+I(9)=27. 27–1=26(Z)

I(9)+G(7)=16. 16–1=15(O)

G(7)+H(8)=15. 15–1=14(N)

H(8)+T(20)=28. 28–1=27=1(mod26)(A)

T(20)+S(19)=39. 39–1=38=12(mod 26)(L)

For OUT: O(15)+U(21)=36. 36–1=35=9(mod 26)(I)

U(21)+T(20)=41. 41–1=40=14(mod 26)(N)

89.

(a) 1975 × 36 / 100 = 711 = **Ocean Drive**

(b) Roman numerals: (VIII × III) + X – (IV × V) = (8 × 3) + 10 – (4 × 5) = XIV = **The Rise of Louis**

(c) A Thousand Times (21 Take 1) = 1000 × (21 – 1) = 20000 = **Leagues Under the Sea**

(d) Atomic numbers of missing parts: (Ta × Co) – (Ga × Sm) = (73 × 27) – (31 × 62) = 49

= In = **Diana Jones and the Temple of Doom** (or **Decent Proposal**, etc.)

(e) Hex missing parts: (EC × DE) + FE – DEF – ED = (236 × 222) + 254 – 3567 – 237

= 48842 = BECA = **Use of Him**

90.

MUSTARD (Bustard) fits between BUSTING (Bunting) and WORMS (Can).

Apollonia	Barbet		Barrel	Baseball	Box
DENTIST	BARBER		BEER	CAP	MATCHES
Boxing	Bunting	**Bustard**	Can	Cary	Case
RING	BUSTING	**MUSTARD**	WORMS	GRANT	WINE
Cecilia	Chicken		Cow	Crispin	Dominic
MUSICIAN	WIRE		PARSLEY	SHOEMAKER	ASTRONOMER
Doris	Duck		Flagon	Florian	Football
DAY	SOUP		CIDER	FIREMAN	BOOTS
Goat	Golf		Harrier	Hockey	Homobonus
WILLOW	BAG		CARRIER	STICK	TAILOR
Horse	Hot		Icing	Jar	John
BOX	HEELS		CAKE	JAM	CANDY
Joseph	Leonard		Lucille	Mae	Pack
CARPENTER	PRISONER		BALL	WEST	CIGARETTES
Pat	Pheasant		Pig	Pile	Plover
BACK	PLEASANT		IRON	AGONY	CLOVER
Polo	Quick		Riding	Sally	Sheep
NECK	DRAW		CROP	FIELD	DIP
Sit	Slap		Swallow	Tennis	Tube
FENCE	WRIST		SHALLOW	ELBOW	TOOTHPASTE
Turkey	Turn		Vincent	Vulture	Will
TROT	HEAT		PRICE	CULTURE	HAY

Themes are:

Patron saints: DENTIST/*APOLLONIA*, MUSICIAN/*CECILIA*, SHOEMAKER/*CRISPIN*, ASTRONOMER/*DOMINIC*, FIREMAN/ *FLORIAN*, TAILOR/*HOMOBONUS*, CARPENTER/*JOSEPH*, PRISONER/*LEONARD*

Bird names with one letter garbled: BARBER/*BARBET*, BUSTING/
BUNTING, **MUSTARD/*BUSTARD***, CARRIER/*HARRIER*,
PLEASANT/*PHEASANT*, CLOVER/*PLOVER*, SHALLOW/
SWALLOW, CULTURE/*VULTURE*

X of Y, where X is a container: *BARREL* of BEER, *BOX* of
MATCHES, *CAN* of WORMS, *CASE* of WINE, *FLAGON*
of CIDER, *JAR* of JAM, *PACK* of CIGARETTES, *TUBE* of
TOOTHPASTE

Can be preceded by a sport: *BASEBALL* CAP, *BOXING* RING,
FOOTBALL BOOTS, *GOLF* BAG, *HOCKEY* STICK, *POLO*
NECK, *RIDING* CROP, *TENNIS* ELBOW

Can be preceded by an actor's Christian name: *CARY* GRANT,
DORIS DAY, *JOHN* CANDY, *LUCILLE* BALL, *MAE* WEST,
SALLY FIELD, *VINCENT* PRICE, *WILL* HAY

Can be preceded by a farmyard animal: *CHICKEN* WIRE, *COW*
PARSLEY, *DUCK* SOUP, *GOAT* WILLOW, *HORSE* BOX, *PIG*
IRON, *SHEEP* DIP, *TURKEY* TROT

X on the Y: *HOT* on the HEELS, *ICING* on the CAKE, *PAT* on the
BACK, *PILE* on the AGONY, *QUICK* on the DRAW, *SIT* on the
FENCE, *SLAP* on the WRIST, *TURN* on the HEAT

91.

(a) **6, 9**. Diagonal pattern on typewriter keyboard

(b) **Q**. Alternate letters of 'To be or not to be that is the question'

92.

(a) **105**. The numbers are increased by the number of letters in their
name. 95 + 10 = 105

(b) **26**. The sequence is the numbers 1, 2, 3, ... added to the lengths
of their names. 26 = 20 + 6

(c) **14**. The sequence is the numbers 1, 2, 3, ... minus the lengths of
their names. 14 = 20 − 6

93.

Pseudonyms, pen names and stage names:

(a) Lady Gaga = Stefani Joanne Angelina Germanotta

(b) Mata Hari = Margaretha Geertruida Zelle MacLeod

(c) Anne Rice = Howard Allen Frances O'Brien

(d) Herbert Lom = Herbert Charles Angelo Kuchačevič ze Schluderpacheru

(e) Joan Fontaine = Joan de Beauvoir de Havilland

(f) Jean Reno = Juan Moreno y Herrera-Jiménez

(g) John Wyndham = John Wyndham Parkes Lucas Beynon Harris

(h) Katie Boyle = Caterina Irene Elena Maria Imperiali di Francavilla

(i) Rudolph Valentino = Rodolfo Alfonso Raffaello Pierre Filibert Guglielmi di Valentina d'Antonguolla

(j) Ellery Queen = Frederic Dannay and Manfred Bennington Lee = Daniel Nathan and Emanuel Benjamin Lepofsky

94.

SWINEHERD (Gauguin) fits between ARMFUL (Fulmar) and BERETS (Green).

Accidental		Avocet	Balthasar	Bar	Bass
UNINTENDED		OCTAVE	SIXTEEN	OBSTRUCT	DEEP
Black		Blue	Brown	Brumaire	Camellias
NARCISSUS		LAGOON	JACKIE	MIST	LADY
Cézanne		Chagall	Cotinga	Degas	Do
GARDENER		JUGGLER	COATING	DANCERS	PERFORM
Fa		Flat	Flies	Floréal	Fructidor
FLORIDA		LEVEL	LORD	FLOWERS	FRUIT
Fulmar	**Gauguin**	Green	Jeroboam	La	Lambs
ARMFUL	**SWINEHERD**	BERETS	FOUR	BEHOLD	SILENCE
Magnum		Manet	Me	Messidor	Methuselah
TWO		LUNCHEON	MYSELF	HARVEST	EIGHT

Mohicans		Monet	Moorhen	Natural	Nebuchadnezzar
LAST		RIVER	HORMONE	NORMAL	TWENTY
Orange		Prairial	Purple	Re	Rehoboam
CLOCKWORK		MEADOWS	COLOR	CONCERNING	SIX
Renoir		Rhea	Rose	Salmanazar	Sharp
UMBRELLAS		HEAR	NAME	TWELVE	ACUTE
Shrike		So	Teal	Thermidor	Ti
HIKERS		THEREFORE	LATE	HEAT	TITANIUM
Treble		Triffids	Ventôse	White	Worlds
THREEFOLD		DAY	WIND	CHRISTMAS	WAR

Themes are:

Synonyms of musical terms: UNINTENDED/*ACCIDENTAL*, OBSTRUCT/*BAR*, DEEP/*BASS*, LEVEL/*FLAT*, NORMAL/ *NATURAL*, ACUTE/*SHARP*, THREEFOLD/*TREBLE*

Anagrams of birds: OCTAVE/*AVOCET*, COATING/*COTINGA*, ARMFUL/*FULMAR*, HORMONE/*MOORHEN*, HEAR/*RHEA*, HIKERS/*SHRIKE*, LATE/*TEAL*

Champagne bottle sizes: SIXTEEN/*BALTHASAR*, FOUR/ *JEROBOAM*, TWO/*MAGNUM*, EIGHT/*METHUSELAH*, TWENTY/*NEBUCHADNEZZAR*, SIX/*REHOBOAM*, TWELVE/ *SALMANAZAR*

Associated with a colour in a film title: *BLACK* NARCISSUS, *BLUE* LAGOON, JACKIE *BROWN*, *GREEN* BERETS, CLOCKWORK *ORANGE*, COLOR *PURPLE*, *WHITE* CHISTMAS

French Revolutionary calendar: MIST/*BRUMAIRE*, FLOWERS/ *FLORÉAL*, FRUIT/*FRUCTIDOR*, HARVEST/*MESSIDOR*, MEADOWS/*PRAIRIAL*, HEAT/*THERMIDOR*, WIND/*VENTÔSE*

X of the Y: LADY/*CAMELLIAS*, LORD/*FLIES*, SILENCE/*LAMBS*, LAST/*MOHICANS*, NAME/*ROSE*, DAY/*TRIFFIDS*, WAR/ *WORLDS*

Painting/artist: GARDENER/*CÉZANNE*, JUGGLER/*CHAGALL*, DANCERS/*DEGAS*, **SWINEHERD**/*GAUGUIN*, LUNCHEON/ *MANET*, RIVER/*MONET*, UMBRELLAS/*RENOIR*

Tonic sol-fa: PERFORM/*DO*, FLORIDA/*FA*, BEHOLD/
LA, MYSELF/*ME*, CONCERNING/*RE*, THEREFORE/*SO*,
TITANIUM/*TI*

95.

(a) **Tonic sol-fa**: doOR, reALLY, miSTRESS, faBLED, soLACE, laTENT, tiTHE, doZEN

(b) **Triangular numbers**: 1, 4, 9, <u>16</u>, 25, <u>36</u>, 49, <u>64</u>, 81, <u>100</u>

(c) **The first elements**: batHe, HEaddress, cLIent, proBEd, noBble, patCh, uniNformed, nOose, Faction, beNEfit, doNAting

(d) **The first 2 letters of the planets**: MEdial, VEnose, EAstern, MAchine, JUice, SAusage, URgently, NEsting

(e) **Points of the compass**: doNor, hoNEst, chEat, reSEnt, haSte, anSWer, seWed, ruNWay

(f) **Roman numerals**: paInt, seVer, maXim, tiLed, seCts, caDre, boMbs

(g) **The numbers ONE to TEN** (each beginning with the end letter of the preceding word and ending with the first letter of the following word): ONTO, ONE, EAT, TWO, OUT, THREE, ENUFF, FOUR, ROOF, FIVE, EYES, SIX, XIS, SEVEN, NURSE, EIGHT, TRAIN, NINE, EAST, TEN

(h) **A, C, ..., Y and B, D, ..., Z**: HEaR/HERb, cLOVE/LOVEd, RATe/RAfT, TANg/ThAN, AiM/jAM, PEAk/PEAl, mASTER/ ASTERn, LOoT/pLOT, qAT/rAT, STARLINGs/STARtLING, VALuE/VALvE, NEwT/NExT, yEN/zEN

96.

(a) **Johann Sebastian Bach** (German)

(b) **Bernhard Langer** (German)

(c) **Franz Kafka** (Czech)

(d) **Giuseppe Verdi** (Italian)

(e) **Enrique Iglesias** (Spanish)

(f) **Zayn Malik** (Arabic)

(g) **Giacomo Casanova** (Italian)

(h) **Boris Pasternak** (Russian)

97.

    (a) **Forms a word when letters are reversed** (ANIMAL/ LAMINA, etc.)

    (b) **Rotates cyclically to form a word** (ALARM/MALAR, etc.)

    (c) **Words where C can be replaced by another letter to form a word** (CLONE/ALONE, etc.)

    (d) **Words on which element names are based** (FLOWING/ FLUORINE, etc.)

    (e) **Words in which PP, P, F or FF can be inserted to form a word** (ARISE/APPRISE, etc.)

98.

AMY. The words can be read down columns when the alphabet is written out on different widths:

7

A	B	C	D	E	F	G
H	I	J	K	L	M	N
O	P	Q	R	S	T	U
V	W	X	Y	Z		

8

A	B	C	D	E	F	G	H
I	J	K	L	M	N	O	P
Q	R	S	T	U	V	W	X
Y	Z						

9

A	B	C	D	E	F	G	H	I
J	K	L	M	N	O	P	Q	R
S	T	U	V	W	X	Y	Z	

12

A	B	C	D	E	F	G	H	I	J	K	L
M	N	O	P	Q	R	S	T	U	V	W	X
Y	Z										

99.

The 3 remaining words – linked by the word **LINK** – are (LINK) **ARMS**, **CONNECTION** (= LINK) and **CHAIN** (LINK).

The pairings are:

Leaders: EDEN, MAJOR, THATCHER (UK) and BUSH, CARTER, HOOVER (US)

Self-anagrams: ASSET, SEATS, TESSA (one female name) and ANDREW, WARDEN, WARNED (one male name)

Beatles' songs: BACK, MADONNA, WRITER (first word omitted) and DAY, ELEANOR, YELLOW (second word omitted)

Anagrams: LUMP, MILE, PAGER (fruits) and APE, BANE, KEEL (vegetables)

Pop brothers: MICHAEL, RANDY, TITO (Jackson) and ALAN, JAY, WAYNE (Osmond)

Homophonic endings: CRAIC, SLACK, YAK (-ack as in black) and INDICT, SITE, TIGHT (-ite as in white)

Homphonic numbers: DRY, FEAR, SEX (German) and CEASE, SANK, WHEAT (French)

Animal association: GUINEA, IGNORANT, NAPOLEON (pig) and BOX, CLOTHES, COPENHAGEN (horse)

Countries: AND, NEW, ZEAL (New Zealand) and ETHER, LANDS, THEN (The Netherlands)

Shakespeare quotes: BE, IF, MUSIC (If music be ...) and SHALL WE WHEN (When shall we ...)

100.

(a) Hart to Hart

(b) Eye to Eye

(c) Bridge to Farr

(d) Longwai to Tipperary

(e) Born to Run

(f) Joy to The World

(g) Freedom to Chooz

(h) Much to Difficult

101.

EON is the odd ONE out being an anagram of a number in English; the other words are anagrams of numbers in French or German:

CHAT/*ACHT*, CHESS/*SECHS*, CORNETIST/ *CENT TROIS*, PEST/*SEPT*, RIDE/*DREI*, RIOTS/*TROIS*, SINE/*EINS*, THUNDER/*HUNDERT*, UNFIXED/*DIX NEUF*, ZONE/*ONZE*

102.

CASTLE (Blandings) fits between KID (Billy) and BLUE (Blood).

All's	Antony		Arthur	As	Bachelors
WELL	CLEOPATRA		MILLER	IT	ANONYMOUS
Bell	Billy	**Blandings**	Blood	Bush	Chain
RINGER	KID	**CASTLE**	BLUE	FIRE	LETTER
Conan	Doctor		Drood	Finger	Fist
BARBARIAN	SALLY		EDWIN	BUFFET	FIGHT
Flood	Food		Ford	French	Gear
FLASH	FAST		FOCUS	LEAVE	BOX

Good **COMMON**	Hand **BAG**	Handlebar **MOUSTACHE**	Harrison **FORD**	Hereward **WAKE**
Hood **ROBIN**	Hot **WATER**	John **BAPTIST**	King **JOHN**	Knuckle **DUSTER**
Lily **PINK**	Lincoln **GREEN**	Love's **LOST**	Measure **MEASURE**	Much **NOTHING**
Palm **TREE**	Performing **FLEA**	Piccadilly **JIM**	Polk **A**	Robert **BRUCE**
Roger **DODGER**	Saddle **SORE**	Sinbad **SAILOR**	Spokes **WOMAN**	Stood **UNDER**
Thumb **TACK**	Thumbnail **SKETCH**	Truman **SHOW**	Twelfth **NIGHT**	Tyre **PRESSURE**
Uncle **DYNAMITE**	Washington **POST**	Wheel **BARROW**	Wood **HOLLY**	Wrist **WATCH**

Themes are:

First/last word in the title of a Shakespeare play: *ALL'S* Well That Ends WELL, *ANTONY* and CLEOPATRA, *AS* You Like IT, *KING* JOHN, *LOVE'S* Labours LOST, *MEASURE* for MEASURE, *MUCH* Ado About NOTHING, *TWELFTH* NIGHT

Can be preceded by a US president's surname: *ARTHUR* MILLER, *BUSH* FIRE, *FORD* FOCUS, *HARRISON* FORD, *LINCOLN* GREEN, *POLK*A, *TRUMAN* SHOW, *WASHINGTON* POST

First/last word in the title of a P. G. Wodehouse book: *BACHELORS* ANONYMOUS, ***BLANDINGS* CASTLE**, *DOCTOR* SALLY, *FRENCH* LEAVE, *HOT* WATER, *PERFORMING* FLEA, *PICCADILLY* JIM, *UNCLE* DYNAMITE

Can be preceded by a part of a bicycle: *BELL* RINGER, *CHAIN* LETTER, *GEAR* BOX, *HANDLEBAR* MOUSTACHE, *SADDLE* SORE, *SPOKES*WOMAN, *TYRE* PRESSURE, *WHEEL*BARROW

Can be followed by a word ending -OOD: BLUE *BLOOD*, EDWIN *DROOD*, FLASH *FLOOD*, FAST *FOOD*, COMMON *GOOD*, ROBIN *HOOD*, UNDER*STOOD*, HOLLY*WOOD*

NAME the DESCRIPTOR: *CONAN* the BARBARIAN, *BILLY* the KID, *HEREWARD* the WAKE, *JOHN* the BAPTIST, *LILY* the PINK, *ROBERT* the BRUCE, *ROGER* the DODGER, *SINBAD* the SAILOR

Can be preceded by a part of the hand: *FINGER* BUFFET, *FIST* FIGHT, *HAND*BAG, *KNUCKLE* DUSTER, *PALM* TREE, *THUMB* TACK, *THUMBNAIL* SKETCH, *WRIST*WATCH

103.

Final	Winner
Henman 6-2, 3-6, 6-1	Murray 6-4, 6-7, 6-1
Murray 6-1, 4-6, 6-1	

The scores derive from comparing the respective letters in the name and seeing which comes earlier in the alphabet. The score in the set is 6-(the number of successive later letters). Where the letters are the same the top player wins the tiebreak. Thus for Henman v Gallek GALL precede each of HENM so that is 6-4 to Henman, A precedes E, so that is 6-1 to Gallek, and K precedes N so that is 6-1 to Henman.

104.

I could have lost by 89-70, a difference of **19**.

```
X|1 1|X|1 1|X|1 1|X|1 1|X|1 1|              Me: 70
9 0|0 0|0 0|0 0|0 0|0 0|0 0|0 X|X|X X X     Opponent: 89
```

105.

Londres. Each line refers to a letter which appears in a country's name in English, but not in the native language.

My first is in Finland but not in Suomi	L
My second is in Estonia but not in Eesti	O
My third is in Albania but not in Shqipëria	N
My fourth is in Sweden but not in Sverige	D

My fifth is in Germany but not in Deutschland   R

My sixth is in Iceland but not in Island   E

My last is in Wales but not in Cymru   S

My whole is in England but not in English   LONDRES

106.

OLDEN. The words can be reparsed to give:

(EMA), NATION-AL, GA- MES, SAGE- LY, CHEE- SE, CRET-
IN, CH- ARM, LESS- ON, SET- TEE, TER- RAPIN, G-OLD, EN-
EMA

to complete the circle.

107.

SALT LAKE CITY is missing from the list of the 50 US state
capitals, which are organised in the order in which their states joined
the union. The words in quotes are translations from a foreign
language.

Synonym for last 4 letters:
Finished (D<u>over</u>); 'Town' (Harris<u>burg</u>); Crossing (Hart<u>ford</u>), String
(Con<u>cord</u>); Castle (Frank<u>fort</u>); Warble (Lan<u>sing</u>); Opera (Hono<u>lulu</u>).

County of state capital:
Mercer (Trenton); Anne Arundel (Annapolis); Richland (Columbia);
Ada (Boise); Washington (Montpelier); Shawnee (Topeka); Laramie
(Cheyenne).

Sports team:
Falcons (Atlanta); Celtics (Boston), Predators (Nashville); Colts
(Indianapolis); Kings (Sacramento); Broncos (Denver); Bruins
(Providence).

Synonym for first 4 letters:
Wealthy (<u>Rich</u>mond); 'Scotland' (<u>Alba</u>ny); High (<u>Tall</u>ahassee);
Bargains (<u>Sale</u>m); Cleaner (<u>Char</u>leston); Vehicles (<u>Cars</u>on City);
Month (<u>June</u>au).

Capital is surname of famous person:

Sir Walter (Raleigh); Christopher (Columbus); Dusty (Springfield); James (Madison); Abraham (Lincoln); Otto von (Bismarck); Joaquin (Phoenix).

Indication of first word in a 2-word capital name:

'Loaf' (<u>Baton</u> Rouge); President (<u>Jefferson</u> City); Small (<u>Little</u> Rock); 'Some' (<u>Des</u> Moines); Simon Templar (<u>Saint</u> Paul); Musical (Oklahoma City); Claus (<u>Santa</u> Fe).

Capital is first name of famous person:

Pollock (Jackson); Clift (Montgomery); Bracknell (Augusta); Powers (Austin); Cardin (Pierre); Rubinstein (Helena); Dukakis (Olympia).

108.

(a) **EVENTS.** Words contain letters of One, Two, etc. EVENTS contains letters of SEVEN.

(b) **UNNERVE.** Words begin with anagrams of German numbers UNNE-RVE.

(c) **PAPERBACK.** Words are anagrams of ALPHA, BETA, etc. plus 4 letters. PAPERBACK is an anagram of KAPPA + ERBC.

109.

This refers to chariots of the Roman gods.

Bacchus was drawn by **Panthers**

Ceres was drawn by **Winged Dragons**

Cybele was drawn by **Lions**

Diana was drawn by **Stags**

Juno was drawn by **Peacocks**

Neptune was drawn by **Sea-Horses**

Pluto was drawn by **Black Horses**

Sun was drawn by **Seven Horses**

Venus was drawn by **Doves**

110.

Roy Jenkins is homophonically Roi Jean Quinze (a), Melchester Rovers' best known player was Roy (Race) of the Rovers (b), Bosworth Field was where Richard of York Gave Battle In Vain (ROYGBIV, the rainbow reminder) (c), corduroy (d), the archaeological excavator of Hissarlik/Troy (e), and the Faroe Islands' Eysturoy and Suduroy (the only 2 of the 18 with that ending) (f).

111.

SUPERCALIFRAGILISTICEXPIALIDOCIOUS: super/Kali/ fragile/Liszt/tick/XP/Ali/do/shoes.

112.

The words can all be associated with Characters, Weapons and Rooms in Cluedo. The Characters precede words, the Weapons are synonyms for words, and the Rooms succeed words. The links are:

ALBERT hall	BRIDGE=spanner	scarlett O'HARA
COCKTAIL lounge	CORD=rope	green PARTY
COPYRIGHT library	OBELISK=dagger	mustard PLASTER
FINE dining	PLUMBING=lead piping	peacock THRONE
FITTED kitchen	SIX-GUN=revolver	plum TOMATO
MUTTART conservatory		
PRIVATE study		
STRICTLY ballroom		

Missing are Mrs White (or Dr Orchid from 2016) who did it in the Billiard Room with the Candlestick.

113.

The completed crossword grid contains each letter of the alphabet exactly once:

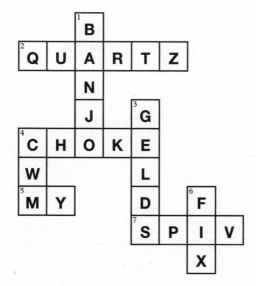

2 Homophone of 'quarts'
4 'OK' in 'Che'
5 Little My is a Moomin character
7 VIPS reversed

114.

18018. The property is that of not containing the letter N. Eighty-eight, achtzig and dix-huit mille dix-huit are the highest in each language.

115.

The solutions to the clues are: EIGHTIES, WRITER, ABOUT, MOLE, WHERE, COUNTRY, BEGINS. These read as a clue to which the answer is the 8-letter word **TOWNSEND**. This is itself a clue to the 8-letter word **SOUTHEND**.

116.

(a) **Anna Ford**

(b) **Pol Pot**

(c) **Idi Amin**

(d) **Kim Wilde**

(e) **Alice Cooper**

(f) **Seb Coe**

(g) **Alan Turing**

(h) **Dave Allen**

(i) **Brad Pitt**

(j) **Ron Ely**

117.

Songs: 50×99

$$\frac{50 \times 99}{2} - 7 = 2468 = \textbf{Motorway}$$

118.

Brad Pitt (whose birthday is 18th December).

```
                  H A P P Y B I R T H D A Y
                H A P P Y B I R T H D A Y
      H A P P Y B I R T H D A Y H A P P Y
        H A P P Y B I R T H D A Y H A P P Y
                  H A P P Y B I R T H D A Y
              H A P P Y B I R T H D A Y
            H A P P Y B I R T H D A Y
            H A P P Y B I R T H D A Y
```

119.

(a) **DOT**, or any word that remains a word with two I's inserted:
FaUNa, bABbLE, OccUR, BIddER, BeReAVE, COffIN,
gRUDgE, ShEATh, iDiOT

(b) **COO**, or any word that remains a word when reversed with
two H's inserted: NIL/LIaNa, DEMO/bOMbED, DAY/cYcAD,
NAME/dEMANd, SPILL/eLLIPSe, UP/PUff, READ/DAggER,
COO/hOOCh

120.

(a) **Is a word when followed by LESS**: isn't, but is a word when
followed by MORE.

(b) **Is a word when preceded by UP**: isn't, but is a word when
preceded by DOWN.

(c) **A Greek letter can be inserted to form a word**: (alphaBET,
DetaIN, DIOxiDE, GRAphiTE, MINOtauR, OpiNION,
PEAnuT, muSLIM, chiTIN, URchiN): can't.

(d) **Can be followed by NUT**: can't, but can be followed by FRUIT.

121.

The list is of London Underground stations beginning with M in
order of Scrabble score, then alphabetically.

MORDEN/9, MILE END/10, MOORGATE/11,
MONUMENT/12, MAIDA VALE/15, MANOR HOUSE/15,
MOOR PARK/16, MANSION HOUSE/17, MARYLEBONE/17,
MILL HILL EAST/17, MARBLE ARCH/19, **MORNINGTON
CRESCENT**/25

122.

(a) Johnny Depp – **Tonto** – Jay Silverheels

(b) Matt Smith – **The Doctor** – **John Hurt** – Winston Smith

(c) Pope Pius XII – **John Gielgud** – **Hobson** (from *Arthur*) – **Helen Mirren** – Queen Elizabeth II

(d) Ewan McGregor – **Obi Wan Kenobi** – **Alec Guinness** – **George Smiley** – **Gary Oldman** – Commissioner Gordon

(e) Steve McQueen – **Thomas Crown** – **Pierce Brosnan** – **James Bond** – **Sean Connery** – **Robin Hood** – Kevin Costner

(f) Denzel Washington – **Captain Bennett Marco** (from *The Manchurian Candidate*) – **Frank Sinatra** – **Danny Ocean** (from *Ocean's 11*) – **George Clooney** – **Batman** – **Michael Keaton** – Betelgeuse

123.

(a) **A TUNE HIDES**: RESURRECTION (M<u>A</u>HLER's 2nd symphony), UNFINISHED (SCHUBER<u>T</u>'s 8th), WAGNER (BR<u>U</u>CKNER's 3rd), ANTARCTICA (VAUGHA<u>N</u>-WILLIAMS' 7th), LITTLE (SCHUB<u>E</u>RT's 6th), 1917 (SHOSTAKOVIC<u>H</u>'s 12th), ORGAN (SA<u>I</u>NT-SAËNS 3rd), ITALIAN (MEN<u>D</u>ELSSOHN's 4th), REFORMATION (MEND<u>E</u>LSSOHN's 5th), SPRING (SCHUMANN's 1st)

(b) **MONARCHY: ENGLAND**: 1553 (year of accession of <u>MA</u>RY I), 1760 (GE<u>O</u>RGE III), 1216 (HE<u>N</u>RY III), 1461 (EDW<u>A</u>RD IV), 1189 (<u>R</u>ICHARD I), 1625 (<u>C</u>HARLES I)), 1660 (<u>C</u>HARLES II), 1413 (HENR<u>Y</u> V): 1558 (<u>E</u>LIZABETH I), 1216 (HE<u>N</u>RY III), 1714 (<u>G</u>EORGE I), 1830 (WIL<u>L</u>IAM IV), 1689 (M<u>A</u>RY II), 1216 (HE<u>N</u>RY III), 1547 (EDWAR<u>D</u> VI)

(c) **LES ROIS FRANCAIS**: 1364 (year of accession of CHAR<u>LE</u>S V), 1380 (CHARL<u>E</u>S VI), 1422 (CHARLE<u>S</u> VII), 922 (<u>R</u>OBERT I), 877 (L<u>O</u>UIS II), 1270 (PH<u>I</u>LIP III), 1422 (CHARL<u>E</u>S VII), 1515 (<u>F</u>RANCIS I), 1589 (HEN<u>R</u>I IV), 893 or 898 (CHAR<u>L</u>ES III), 1574 (HE<u>N</u>RI III), 481 (<u>C</u>LOVIS I), 893 or 898 (CHAR<u>L</u>ES III), 936 (LOU<u>I</u>S IV), 986 (LOUI<u>S</u> V)

(d) **DAYS OF THE YEAR**: 1/12 (DECEMBER 1ˢᵗ), 1/8 (AUGUST 1ˢᵗ), 7/1 (JANUARY 7ᵗʰ), 1/9 (SEPTEMBER 1ˢᵗ), 4/10 (OCTOBER 4ᵗʰ), 1/2 (FEBRUARY 1ˢᵗ), 4/9 (SEPTEMBER 4ᵗʰ), 5/3 (MARCH 5ᵗʰ), 7/11 (NOVEMBER 7ᵗʰ), 4/7 (JULY 4ᵗʰ), 4/6 (JUNE 4ᵗʰ), 2/5 (MAY 2ⁿᵈ), 3/3 (MARCH 3ʳᵈ)

(e) **COUNT DRACULA**: 1, 3/C, 2, 15/O, 3, 21/U, 4, 14/N, 5, 20/T, 6, 4/D, 7, 18/R, 8, 1/A, 9, 3/C, 10, 21/U, 11, 12/L, 12, 1/A

124.

Film titles 'containing' chemical elements (or homophones), at the front (F) or rear (R):

6 (Carbon)	Copy	F	Carbon Copy
10 (Neon)	The Virgin	R	The Virginian
14 (Silicon)	Valley	F	Silicon Valley
26 (Iron)	Pumping	R	Pumping Iron
29 (Copper)	Spare a	R	Spare a Copper
30 (Zinc)	Monster	R	Monsters Inc.
33 (Arsenic)	And Old Lace	F	Arsenic and Old Lace
47 (Silver)	Streak	F	Silver Streak
78 (Platinum)	Blonde	F	Platinum Blonde
79 (Gold)	Finger	F	Goldfinger
80 (Mercury)	Rising	F	Mercury Rising
82 (Lead)	Belly	F	Leadbelly
86 (Radon)	Entebbe	F	Raid on Entebbe

Alternative answers exist for some rows.

125.

(a) **6.** The sequence is the number of letters in the names of months. August has 6 letters.

(b) **8.** The sequence is the number of letters in each word of 'To be or not to be that is the question'. 'Question' has 8 letters.

(c) **3.** The sequence is the length of the roman numbers. XVI has length 3.

126.

YULETIDE GREETINGS. O3 = 3rd letter of O/OXY̲GEN, K8 = 8th letter of K/POTASSIU̲M . . . P4 = 4th letter of P/PHOS̲PHORUS.

127.

(37, 38, 39).

(910, 911, 912) = NINE HUNDRED AND TEN̲/N̲INE HUNDRED AND ELEVEN̲/N̲INE HUNDRED AND TWELVE

(511, 512, 513) = FÜNFHUNDERTEL F̲/ F̲ÜNFHUNDERTZWÖL F̲/F̲ÜNFHUNDERTDREIZEHN

(37, 38, 39) = TRENTE-SEP T̲/T̲RENTE-HUI T̲/T̲RENTE-NEUF

These are the smallest triples of consecutive numbers in each language where the last letter of a number is the first letter of its successor.

128.

Opening lines of pop songs with one-word titles starting A–Z.

Angels Robbie Williams	I sit and wait, does an angel contemplate my fate
Blockbuster The Sweet	You better beware, you better take care
Cars Gary Numan	Here in my car I feel safest of all
Delilah Tom Jones	I saw the light on the night that I passed by her window
Easy The Commodores	Know it sounds funny but I just can't stand the pain
Fever Peggy Lee/Elvis Presley	Never know how much I love you, never know how much I care
Ghostbusters Ray Parker Jr.	If there's something strange in your neighborhood, who you gonna call
Hallelujah Leonard Cohen, etc.	I've heard there was a secret chord that David played and it pleased the Lord
Imagine John Lennon	Imagine there's no heaven, it's easy if you try
Jolene Dolly Parton	Jolene, Jolene, Jolene, Jolene, I'm begging of you please don't take my man

Kayleigh Marillion	Do you remember, chalk hearts melting on a playground wall
Lola The Kinks	I met her in a club down in old Soho
Mandy Barry Manilow	I remember all my life, raining down as cold as ice
Nineteen Paul Hardcastle	In 1965 Vietnam seemed like just another foreign war
One U2	Is it getting better, or do you feel the same
Panic The Smiths	Panic on the streets of London, panic on the streets of Birmingham
Question The Moody Blues	Why do we never get an answer when we're knocking at the door
Relax Frankie Goes To Hollywood	Relax, don't do it, when you want to go to it
Sailing Rod Stewart	I am sailing, I am sailing, home again 'cross the sea
Thriller Michael Jackson	It's close to midnight and something evil's lurking in the dark
Unbelievable EMF	You burden me with your questions, you'd have me tell no lies
Valerie The Zutons/Amy Winehouse	Well sometimes I go out by myself, and I look across the water
Waterloo Abba	My my, at Waterloo Napoleon did surrender
Xanadu Olivia Newton-John	A place where nobody dared to go, the love that we came to know
Yesterday The Beatles	Yesterday, all my troubles seemed so far away
Zombie The Cranberries	Another head hangs lowly, child is slowly taken

129.

(a) $1138 - (101 \times 5) = 633 = $ **Squadron**

(b) $2 + 5 = 7 = $ **Samurai** (or **The Magnificent**, etc.)

(c) $29 + (84 \times 4) = 365 = $ **Nights in Hollywood**

(d) $2001 + 98 - 99 = 2000 = $ **Death Race**

(e) $22 \times (451 + 1) / 100 = 99.44 = $ **% Dead**

130.

    (a) **USAGE**

    (b) **STATE RECOGNISED CHURCH**

From (b) the cipher alphabet is:
STAERCOGNIDHUBFJKLMPQVWXYZ. A word of length n is enciphered by replacing each letter by the letter n to its right in the cipher alphabet. Thus 'IN ECCLESIASTICAL USAGE THE TERM ESTABLISHMENT IS APPLIED TO' enciphers to the text in the question. As it is of length 26, ANTIDISESTABLISHMENTARIANS enciphers to itself.

131.

    (a) **Each letter is replaced by the letter that is its value** (A=1, B=2, ..., Z=26) further ahead in the quotation. Thus S is replaced by letter 1+19 =20=L, E by 2+5=7=E, A by 3+1=4=S and so on.

    (b) **NO, IT IS A VERY INTERESTING NUMBER;** IT IS THE SMALLEST NUMBER EXPRESSIBLE AS A SUM OF CUBES IN TWO DIFFERENT WAYS.

    (c) **HE KNOWS DEATH TO THE BONE –** MAN HAS CREATED DEATH.

    (d) **HAVE YOU HEARD? IT'S IN THE STARS,** NEXT JULY WE COLLIDE WITH MARS.

    (e) **ONE RING TO RULE THEM ALL, ONE RING TO FIND THEM,** ONE RING TO BRING THEM ALL AND IN THE DARKNESS BIND THEM.

    (f) **IT WAS A BRIGHT COLD DAY IN APRIL,** AND THE CLOCKS WERE STRIKING THIRTEEN.

    (g) **BUT, SOFT! WHAT LIGHT THROUGH YONDER WINDOW BREAKS?** IT IS THE EAST, AND JULIET IS THE SUN.

132.

1: **THE** word in the set of one is THE.

2 words that form to make a pair: COUP, LED

3 words forming the motto of The Three Musketeers: ALL, FOR, ONE (and ONE, FOR, ALL!)

4 colours of the horses of the 4 Horsemen of the Apocalypse: BLACK, PALE, RED, WHITE

5 members of sets of 5: BABY (Spice Girls), BORODIN (The 5), KESH (The 5 K's), PRAYER (the 5 tenets of Islam), SIGHT (the 5 senses)

6 geese (a-laying!): BERRY, BRENT, CANADA, GOLDEN, MOTHER, SNOW

7 The Magnificent Seven: AUGUST, FINE, GRAND, GREAT, MAJESTIC, NOBLE, SUPERB

8 eight-letter words, the first and last letters spelling OCTUPLES:

OPERATIC
CHESTNUT
TIRAMISU
UNDERLAP
PORTUGAL
LACROSSE
ELEVATES
SOMBRERO

9 words linked with the 9 planets: (Mercury) THERMOMETER, WOMEN (are from Venus), RARE (Earth), (Mars) BAR, ZEUS (Jupiter), (Saturn) V, TITANIA (moon of Uranus), (Neptune's) TRIDENT, (Pluto's master) MICKEY

10 words containing TEN, all at different offsets within the word: TENNIS, STENCH, BUTENE, LISTEN, BRITTEN, LIECHTENSTEIN, FORGOTTEN, SUPERINTENDENT, MOUNTBATTEN, OMNICOMPETENCE

133.

The answers to a–h are eight trigraphs, using 24 of the 26 letters of the alphabet, with N and Y left over.

- (a) QPR
- (b) DMZ
- (c) KGB
- (d) XTC
- (e) VHF
- (f) JLS
- (g) AWE
- (h) IOU
- (i) NY = **New York**

134.

- (a) **CURDLE**. In the others the letter score (A=1, B=2, ..., Z=26) of letters 4–6 is twice that of letters 1–3 (e.g. EDE=5+4+5=14, ACC=1+3+3=7); for CURDLE it's the opposite (DLE=4+12+5=21, CUR=3+21+18=42).

- (b) **FROZEN**. In the others the Scrabble score of letters 4–6 is half that of letters 1–3 (e.g. IND=4, BEH=8); for FROZEN it's the opposite (ZEN=12, FRO=6).

135.

XI. The table is the alphabet written out spirally from the centre:

U	V	W	X	Y
T	G	H	I	J
S	F	A	B	K
R	E	D	C	L
Q	P	O	N	M

136.

YULE (Log) fits between CORPORAL (Lance) and PETER (Mary).

Battle	Bayonet	Bog	Bottle	Boy	Burning
LAST	FITTING	PEAT	MESSAGE	HORSE	LOVE

Candle	Candlestick Maker	Caspian	Chair	China Shop	Dog
BELL	BUTCHER	PRINCE	SILVER	BULL	TOY

Favour	Glove	Gog	Handsome	Harry	Hat
CALL	FOX	MA	TALL	TOM	OLD

Heartbreak	Hog	Hound	Jacket	Jailhouse	Lance	**Log**
HOTEL	GROUND	DOG	DONKEY	ROCK	CORPORAL	**YULE**

Mary	Match	Million	Nephew	Nog	Pants
PETER	GAME	CHANCE	MAGICIAN'S	EGG	HOT

Pear Tree	Poke	Rifle	Sabre	Shirt	Shoe
PARTRIDGE	PIG	RANGE	TOOTH	HAIR	PLATFORM

Shotgun	Sinker	Socks	Spear	Suspicious	Sword
WEDDING	HOOK	BOBBY	MINT	MINDS	FISH

Teacup	Treader	Wardrobe	Way	Wooden	Zog
STORM	VOYAGE	LION	DOWN	HEART	KING

Themes are:

Chronicles of Narnia, first/last word ignoring 'the': The LAST *BATTLE*, The HORSE and His *BOY*, PRINCE *CASPIAN*, The SILVER *CHAIR*, The MAGICIAN'S *NEPHEW*, The VOYAGE of the 'Dawn *TREADER*', The LION, the Witch and the *WARDROBE*

Can be preceded by a weapon: *BAYONET*-FITTING, *LANCE* CORPORAL, *RIFLE* RANGE, *SABRE*-TOOTH, *SHOTGUN* WEDDING, *SPEAR*MINT, *SWORD*FISH

Can be followed by a 3-letter word ending -OG: PEAT *BOG*, TOY *DOG*, MA*GOG*, GROUND*HOG*, **YULE LOG**, EGG*NOG*, KING *ZOG*

X in a Y: MESSAGE in a *BOTTLE*, BULL in a *CHINA SHOP*, CALL in a *FAVOUR*, CHANCE in a *MILLION*, PARTRIDGE in a *PEAR TREE*, PIG in a *POKE*, STORM in a *TEACUP*

First/last word of Elvis song: *BURNING* LOVE, *HEARTBREAK* HOTEL, *HOUND* DOG, *JAILHOUSE* ROCK, *SUSPICIOUS* MINDS, *WAY* DOWN, *WOODEN* HEART

X,Y and Z: BELL, BOOK and *CANDLE*; BUTCHER, BAKER and *CANDLESTICK MAKER*; TALL, DARK and *HANDSOME*; TOM, DICK and *HARRY*; PETER, PAUL and *MARY*; GAME, SET and *MATCH*; HOOK LINE and *SINKER*

Can be followed by a garment: FOX*GLOVE*, OLD *HAT*, DONKEY *JACKET*, HOT*PANTS*, HAIR *SHIRT*, PLATFORM *SHOE*, BOBBY*SOCKS*

137.

(a) **NATASHA**. Girls' names beginning with the same letters as MERCURY, VENUS, etc., and the same length, and with no other letters in the same places.

(b) **ACCOUNTANT**. Words contain the first letters of One, Two; Two, Three; Three, Four; ... in the right places. ACCOUNTANT has N and T in places 9 and 10.

(c) **JOG**. Words rhyme with THE, QUICK, BROWN, FOX, etc., and the complete sequence contains all the letters of the alphabet, like the original.

138.

ANIMAL (Tier) fits between PIGEON (Squab) and CAT (Tom).

Belette	Bitch	Buck	Calf	Cerf
WEASEL	WOLF	ANTELOPE	BISON	DEER
Chameau	Chick	Cob	Cock	Colony
CAMEL	HERON	SWAN	OSTRICH	TERMITE
Crapaud	Crash	Creep	Cria	Doe
TOAD	RHINOCEROS	TORTOISE	ALPACA	RAT
Dog	Drake	Drey	Elver	Ewe
COYOTE	DUCK	SQUIRREL	EEL	SHEEP
Eyas	Eyrie	Form	Gander	Geier
HAWK	EAGLE	HARE	GOOSE	VULTURE
Hen	Hob	Holt	Igel	Jenny
LOBSTER	POLECAT	OTTER	HEDGEHOG	ASS

Joey KANGAROO	Kaninchen RABBIT	Krokodil CROCODILE	Lachs SALMON	Leap LEOPARD
Lodge BEAVER	Manchot PENGUIN	Mare HORSE	Maus MOUSE	Murder CROW
Nest HORNET	Parliament OWL	Pod WHALE	Pride LION	Pup WALRUS
Renne REINDEER	Requin SHARK	Rookery SEAL	Schimpanse CHIMPANZEE	Sett BADGER
Singe MONKEY	Sow BEAR	Squab PIGEON	**Tier** **ANIMAL** Tom CAT	Vixen FOX

Themes are:

Animal/French: WEASEL/*BELLETTE*, DEER/*CERF*, CAMEL/ *CHAMEAU*, TOAD/*CRAPAUD*, PENGUIN/*MANCHOT*, REINDEER/*RENNE*, SHARK/*REQUIN*, MONKEY/*SINGE*

Animal/female: WOLF/*BITCH*, RAT/*DOE*, SHEEP/*EWE*, LOBSTER/*HEN*, ASS/*JENNY*, HORSE/*MARE*, BEAR/*SOW*, FOX/*VIXEN*

Animal/male: ANTELOPE/*BUCK*, SWAN/*COB*, OSTRICH/*COCK*, COYOTE/*DOG*, DUCK/*DRAKE*, GOOSE/*GANDER*, POLECAT/ *HOB*, CAT/*TOM*

Animal/young: BISON/*CALF*, HERON/*CHICK*, ALPACA/*CRIA*, EEL/*ELVER*, HAWK/*EYAS*, KANGAROO/*JOEY*, WALRUS/*PUP*, PIGEON/*SQUAB*

Animal/collective noun: TERMITE/*COLONY*, RHINOCEROS/ *CRASH*, TORTOISE/*CREEP*, LEOPARD/*LEAP*, CROW/*MURDER*, OWL/*PARLIAMENT*, WHALE/*POD*, LION/*PRIDE*

Animal/habitat: SQUIRREL/*DREY*, EAGLE/*EYRIE*, HARE/*FORM*, OTTER/*HOLT*, BEAVER/*LODGE*, HORNET/*NEST*, SEAL/ *ROOKERY*, BADGER/*SETT*

Animal/German: VULTURE/*GEIER*, HEDGEHOG/*IGEL*, RABBIT/ *KANINCHEN*, CROCODILE/*KROKODIL*, SALMON/*LACHS*, MOUSE/*MAUS*, CHIMPANZEE/*SHIMPANSE*, **ANIMAL**/***TIER***

139.

9. There are 9 words in the sentence asked about, 22 words in the kwestion and 6 in the self-referential sentence.

140.

Single-word album titles, and **selected tracks** from them:

Arrival (ABBA)	Dancing Queen, Knowing Me Knowing You, Money Money Money
Britney (Britney Spears)	I'm A Slave 4 U, Overprotected, I'm Not A Girl Not Yet A Woman, I Love Rock'n'Roll
Caribou (Elton John)	The Bitch Is Back, You're So Static, Don't Let The Sun Go Down On Me
Dare (Human League)	Open Your Heart, The Sound Of The Crowd, Love Action (I Believe In Love), Don't You Want Me
Elephant (The White Stripes)	Seven Nation Army, I Just Don't Know What To Do With Myself, The Hardest Button To Button
Faith (George Michael)	Faith, Father Figure, I Want Your Sex (Parts 1 & 2)
Graceland (Paul Simon)	The Boy In The Bubble, Graceland [or Gumboots], Diamonds On The Soles Of Her Shoes, You Can Call Me Al
Heroes (David Bowie)	Heroes, Sons Of The Silent Age, V-2 Schneider, The Secret Life Of Arabia
Imagine (John Lennon)	Imagine, Jealous Guy, Gimme Some Truth, How Do You Sleep?
Jollification (Lightning Seeds)	Perfect, Lucky You, Change, Marvellous
Kick (INXS)	New Sensation, Devil Inside, Need You Tonight, Mystify [or Mediate]
Legend (Bob Marley)	No Woman No Cry, Could You Be Loved, Buffalo Soldier, Redemption Song, Exodus

Metallica (Metallica)	Enter Sandman, Sad But True, Nothing Else Matters, Of Wolf And Man
Nevermind (Nirvana)	Smells Like Teen Spirit, In Bloom, Come As You Are, Lithium
Ommadawn (Mike Oldfield)	Ommadawn Part One, Ommadawn Part Two, On Horseback
Parklife (Blur)	Girls & Boys, Parklife, To The End, This Is A Low
Quadrophenia (The Who)	The Real Me, I'm One, Bell Boy, Love Reign O'er Me
Revolver (The Beatles)	Taxman, Eleanor Rigby, Here, There And Everywhere, Yellow Submarine, Tomorrow Never Knows
Spice (Spice Girls)	Wannabe, Say You'll Be There, 2 Become 1, Who Do You Think You Are
Thriller (Michael Jackson)	Wanna Be Startin' Something, Thriller, Beat It, Billie Jean
Ummagumma (Pink Floyd)	Careful With That Axe Eugene, Set The Controls For The Heart Of The Sun, Several Species Of Small Furry Animals Gathered Together In A Cave And Grooving With A Pict
Violator (Depeche Mode)	World In My Eyes, Personal Jesus, Enjoy The Silence, Policy Of Truth
War (U2)	Sunday Bloody Sunday, New Year's Day, Two Hearts Beat As One
X (Kylie Minogue)	2 Hearts, In My Arms, The One, All I See, Wow
Youthquake (Dead or Alive)	You Spin Me Round (Like a Record), In Too Deep, Lover Come Back To Me
Zoolook (Jean-Michel Jarre)	Ethnicolor, Diva, Zoolook, Wooloomooloo, Zoolookologie, Blah-Blah Café

141.

17. Ignoring case, there are 17 distinct letters in 'How many are to be found in this sentence?' The other numbers similarly reflect a sentence and the kwestion.

142.

(a) **NETWORK**. The numbers ONE, TWO, ..., TEN are spelled out by moving vertically and horizontally through the letters starting with <u>O</u>:

```
OoliTHs
NETWORk
assUREs
proOFEd
dislIke
sleEVes
eXISted
iSolate
kETchup
eVENINg
blIGHTy
```

(b) **HELIPAD** (for example). The symbols of the first 18 elements are spelled out by moving vertically and horizontally through the letters starting with <u>H</u>:

```
Hundred
HELIpad
plEBian
cONCert
aFfable
ANEmone
MustARd
GALilee
inSeCts
chIPSet
```

(c) **EVOLVED** (for example). The quote 'IF MUSIC BE THE FOOD OF LOVE PLAY ON' is spelled out by moving vertically and horizontally through the letters starting with I̲:

```
ofFIcer
SUMmary
IcEFloe
CaHOOts
BETiDes
comFOrt
EVOLved
PLinths
bAYONet
```

(d) **KNIGHT**. The words KNIGHT KNIGHT KNIGHT are spelled out using Knight's moves starting with K̲:

```
ANKLES .NK...
BUNKER ..NK..
KNIGHT KNIGHT
WITHER .ITH..
WEIGHS ..IGH.
OUTAGE ..T.G.
```

(e) **ANEMONE**. It enables N̲ORTH, S̲OUTH, E̲AST and W̲EST to be spelled out appropriately:

```
heaRTHs
SEnOrAS
TWiNSET
anemOne
acHTUng
```

(f) **324**. All the numbers are squares and 0, 1, ..., 9 can be traced by moving vertically and horizontally through the digits:

```
016 01.
324 32.
484 4..
529 5..
676 67.
289 .89
```

143.

    (a) It's a Wonderful Life

    (b) Miracle on Thirty-Fourth Street

    (c) Holiday Inn

    (d) Dr Seuss' How the Grinch Stole Christmas

    (e) Beyond Tomorrow

    (f) The Muppet Christmas Carol

    (g) Love Actually

    (h) The Nightmare Before Christmas

    (i) The Snowman

    (j) Jingle All the Way

    (k) The Polar Express

    (l) The Fourth Wise Man

144.

2. The sentence contains – ignoring case – two H's.

145.

Anagrams of members of the **Rolling Stones**:

Charlie Watts, Keith Richards, Bill Wyman, Brian Jones, Mick Jagger

146.

1984.

    (a) **EG** can be inserted to form a word (BegIN, BOWLegS, LegATE, LIegE, RegAINED) but not in EXAMPLE.

    (b) **OR**. The lengths of the words are decreasing uniformly, and the first/last letters (alternately) form sequences.

W	S	U	Q	S	O
T	V	R	T	P	R

(c) Synonyms ending in **EW**: GUSTED/BLEW, MASTICATE/
CHEW, EIGHT/CREW, AWARE/KNEW, MODERN/NEW,
STITCH/SEW, RAGOUT/STEW, HURLED/THREW

(d) Associated with a 2-word phrase, both words ending in **LL**:
BAD FEELING/ILL WILL, TARANTINO FILM/KILL BILL,
LONDON THOROUGHFARE/PALL MALL, CONFUSED/
PELL MELL, MUSTER/ROLL CALL

The 4×2 grid is therefore: EG

OR

EW

LL

The name **GEORGE ORWELL** can be read by moving vertically
and horizontally through the letters. He wrote **1984**.

147.

BYATT(/AS/33). The associations are NAME/
INITIALS=ELEMENT SYMBOL/ATOMIC NUMBER:

BATES/HE/2, BURROUGHS/ER/68, CLARKE/AC/89,
GARDNER/ES/99, HEINLEIN/RA/88, JAMES/PD/46, LEWIS/
CS/55, SHELLEY/PB/82, WELLS/HG/80, WOOLF/V/23

148.

EIGHTY ONE has **9** letters, $9^2=81$, ONE HUNDRED has **10** letters
$10^2=100$, FIVE HUNDRED AND SEVENTY SIX has **24** letters,
$24^2=576$.

149.

STUNNER/LULU. The others are associated with words/phrases
of the form ?A?A or ?O?O: ARTFORM/DADA, AVERAGE/SOSO,
BIRD/DODO, CLOWN/COCO, DITCH/HAHA, DOG/TOTO,
EXECUTIONER/KOKO, FATHER/PAPA, FORBIDDANCE/
NONO, GOODBYE/TATA, GRANDMOTHER/NANA, MOTHER/
MAMA, SENILE/GAGA, STUNNER, TOY/YOYO

150.

(a) **All appear in the name of A MUSE** (CALLiope, terpsiCHORE, polyHYMNia, melpOMENe, uRANia, eRATo): doesn't.

(b) **All appear in the name of one of the uNITed states** (alASKa, orEGOn, new hampsHIRE, deLAWare, misSOURi, kenTUCKy): doesn't.

151.

OVERT.

1: OVERT(-ONE)

2: Ronnies, Ronnie SCOTT and Ronnie WOOD

3: CHEERs, KINGs, SISTERs

The Sign of Four's 4: Dost AKBAR, Abdullah KHAN, Mohamet SINGH, Jonathan SMALL

Can be preceded by the Famous 5's names: Anne BOLEYN, George BUSH, Julian CALENDAR, Dick CHENEY, Timmy TIPTOES

6: Characters in Search of an Author (by Pirandello): AN, AUTHOR, CHARACTERS, IN, OF, SEARCH

From a set of 7: ANTHONY (Champions of Christendom), ATLANTIC (Seas), BIAS (Sages), COLIN (Secret Seven), COLOSSUS (Wonders), GREEN (Rainbow), PRIDE (Deadly Sins)

Rhymes with 8/EIGHT: BAIT, EYOT, FETE, FREIGHT, KAYTE, LATE, STRAIGHT, THWAITE

Contains a contiguous anagram of NINE: (3×ENNI, 3×INNE, 3×ENIN): DINNER, JENNINGS, LENIN, OPENING, PENINSULAR, PFENNIG, SINNED, SPINNEY, TENNIS

10 Lords (a-leaping): lord CHANCELLOR, GOOD lord, JACK lord, lord JIM, lord KNOWS, LANDlord, LAW lord, lord MAYOR, lordSHIP, TIME lord

152.

08° 42' N	05° 58' W	=	Boron
30° 10' S	23° 08' E	=	Sodium
32° 04' N	100° 41' W	=	Silver
32° 16' N	98° 50' W	=	Carbon
34° 33' N	96° 59' W	=	Sulphur
35° 25' N	114° 11' W	=	Chloride
36° 40' N	115° 59' W	=	Mercury
41° 58' N	02° 47' E	=	Salt
44° 21' N	103° 46' W	=	Lead
47° 24' N	79° 41' W	=	Cobalt
58° 59' N	161° 48' W	=	Platinum
61° 57' N	128° 12' W	=	Tungsten

All are elements, except for **Salt** and **Chloride**.

Salt = Sodium + Chloride.

153.

(a) **23.** The sequence is the first numbers with E in the 1st, 2nd, 3rd position, etc. Twenty-three is the first to have E in the 10th position.

(b) **47.** The sequence is the first numbers with N in the 1st, 2nd, 3rd position, etc. Forty-seven is the first to have N in the 10th position.

(c) **5000.** These are the numbers with no repeated letters in their names.

154.

(a) **Van Halen**

(b) **Duran Duran**

(c) **Yes**

(d) **Bee Gees**

(e) **Abba**

(f) **AC/DC**

(g) **REM**

(h) **Linkin Park**

(i) **New Order**

(j) **Bon Jovi**

(k) **The Real Thing**

(l) **Sum 41**

155.

18° 20' N	74° 22' E	=	SUPA
50° 57' N	01° 51' E	=	CALAIS
41° 31' N	00° 21' E	=	FRAGA
55° 01' N	08° 26' E	=	LIST
53° 34' N	08° 16' E	=	STICK
50° 40' N	03° 27' W	=	EXE
13° 24' N	98° 31' E	=	PE
37° 48' N	01° 34' W	=	ALEDO
32° 10' N	48° 15' E	=	SHUSH

156.

The smallest (we could create) is of **size 25**:

A	S	H	F	G
B	S	L	U	R
H	N	E	M	D
A	Y	E	P	O
P	P	Z	Y	C

157.

SOFT (soap) fits between DIVER (sky) and MUSIC (soul).

Backwards	Basin	Bath	Beethoven	Blair
BEND	PUDDING	EARLY	ROLL	WITCH
Board	Bowler	Brown	Cry	Dry
MAN	SPIN	WINDSOR	BABY	DOCK
Eared	Eyed	Fist	Fly	Footed
DOG	CROSS	HAND	WEIGHT	FLAT
Fry	Gladstone	Going	Handed	Headed
UP	BAG	POSTAL	EVEN	PIG
Heath	Heels	Interesting	Keeper	Kneed
ROW	HEAD	TIMES	WICKET	KNOCK

Leg	Legged	Loo		Major	Making
SQUARE	BOW	WATER		GENERAL	MONEY
Man	Matter	Mirror		Monstrous	Night
THIRD	MIND	DAILY		REGIMENT	WATCH
North	Nosed	Off/On		Out	Plug
POLE	HARD	MID		RUN	VOLCANIC
Ply	Point	Salisbury		Sands	Screen
MOUTH	COVER	PLAIN		GRANGE	SIGHT
Shower	Shy	Sky	**Soap**	Soul	Taking
APRIL	LOCK	DIVER	**SOFT**	MUSIC	NO
Tap	Wellington	Witches		Wry	Wyrd
SPINAL	BOOT	ABROAD		NECK	SISTERS

Themes are:

X over Y: BEND over *BACKWARDS*, ROLL over *BEETHOVEN*, MAN over *BOARD*, HAND over *FIST*, HEAD over *HEELS*, MIND over *MATTER*, GRANGE-over-*SANDS*, NO over *TAKING*

Can be followed by something found in a bathroom: PUDDING *BASIN*, EARLY *BATH*, WATER*LOO*, DAILY *MIRROR*, VOLCANIC *PLUG*, APRIL *SHOWER*, **SOFT** **SOAP**, SPINAL *TAP*

Can be preceded by a Prime Minister's name: *BLAIR* WITCH, *BROWN* WINDSOR, *GLADSTONE* BAG, *HEATH* ROW, *MAJOR* GENERAL, *NORTH* POLE, *SALISBURY* PLAIN, *WELLINGTON* BOOT

Can be followed by a word to give a cricketing term: SPIN *BOWLER*, WICKET *KEEPER*, SQUARE *LEG*, THIRD *MAN*, MID *OFF/ON*, RUN *OUT*, COVER *POINT*, SIGHT *SCREEN*

Can be preceded by a 3-letter word ending -Y: *CRY* BABY, *DRY* DOCK, *FLY* WEIGHT, *FRY* UP, *PLY* MOUTH, *SHY* LOCK, *SKY* DIVER, *WRY* NECK

Can be followed by the past participle of a part of the body: DOG *EARED*, CROSS *EYED*, FLAT *FOOTED*, EVEN *HANDED*, PIG *HEADED*, KNOCK *KNEED*, BOW *LEGGED*, HARD *NOSED*

Can be preceded by a word to give a Terry Pratchett book:
GOING POSTAL, *INTERESTING* TIMES, *MAKING* MONEY,
MONSTROUS REGIMENT, *NIGHT* WATCH, *SOUL* MUSIC,
WITCHES ABROAD, *WYRD* SISTERS

158.

SINE.

Anagram of 1 in German (EINS): SINE

The mummy of all puns: 2/TWO TEN CARMEN

3 Men in a Boat: GEORGE, HARRIS, JEROME

Can be preceded by 4/FOUR: FREEDOMS, HORSEMEN,
POSTER, SEASONS

Synonym rhymes with 5/FIVE and initial letters spell out
FIVES: FLOURISH/THRIVE, IMPEL/DRIVE, VITAL/ALIVE,
ENGINEER/CONTRIVE, STRUGGLE/STRIVE

6 of the best (synonymously): FINEST, FOREMOST, LEADING,
OPTIMAL, SUPREME, UNSURPASSED

7th in a series: ETA (Greek alphabet), GOLF (NATO phonetic),
JULY (month), MARYLAND (admitted to the Union), NITROGEN
(element), PARIS (Summer Olympics), VIOLET (rainbow)

8 maid(en)s a-milking: CHAMBERmaid, IRON maiden, maid
MARIAN, OLD maid, maiden OVER, PARLOURmaid, maiden
SPEECH, maiden VOYAGE

Words that are the same length as the 9 Muses and have the first
3 letters the same: CALMNESS, CLIP, ERASE, EUTROPY,
MELBOURNE, POLITICIAN, TERRITORIAL, THAMES,
URALIC

Contains a X (10 in Roman numerals): BOXER, EXTRA,
FOXTROT, JUXTAPOSE, NEXT, OXEN, SAXOPHONE,
SUSSEX, TEXAS, WAXY

159.

    (a) **FIBONACCI**: (0) 1, 1, 2, 3, 5, 8, 13, 21 (the sequence itself)

    (b) **DICKENS**: (0) 4-9-3-11-5-14-19 (1=A, 2=B, ..., 26=Z)

    (c) **GEORGE BUSH**: (0) 4-3-6-7-4-3-2-8-7-4 (telephone keypad, 2=ABC, 3-DEF, ..., 9=WXYZ)

    (d) **BILL BLOGGS**: as read upside down on a calculator, 0 not used

    (e) **NO ONE OF NOTE**: first digits of 0/<u>N</u>OUGHT, 1/<u>O</u>NE, ..., 8/ <u>E</u>IGHT

160.

OLIVIA. Each word can become a country by adding a letter at the beginning, and these letters are in alphabetical order.

B-OLIVIA, C-HAD, F-INLAND, I-RAN, O-MAN, S-PAIN

161.

    (a) **Snow White's dwarfs**: BashfuL, DoC, DopeY, GrumpY, HappY, SleepY, SneezY

    (b) **The Gospels**: MattheW, MarK, LukE, JohN

    (c) **The Muses**: CalliopE, CliO, EratO, EuterpE, MelpomenE, PolyhymniA, TerpsichorE, ThaliA, UraniA

    (d) **The Fellowship of the Ring**: AragorN, BoromiR, FrodO, GandalF, GimlI, LegolaS, MerrY, PippiN, SaM

    (e) **The 'Four' Musketeers**: AramiS, AthoS, PorthoS, D'artagnaN

162.

Henry Miller's lover   = (Anaïs) NIN
Bed artist   = (Tracey) EMIN
American naval vessel = USS
Number nine   = IX

NINEMINUSSIX = NINE MINUS SIX = THREE = **Thereabout (5)**

163.

LARGE HADRON COLLIDER. With A=1, B=2, ..., Z=26:

```
LARGE HADRON COLLIDER plain+
LHCLH CLHCLH CLHCLHCL key=
XIUSM KMLUAV FATOULHD cipher
```

164.

C. Auguste Dupin was a fictional detective created by <u>Edgar</u> Allen Poe (a); 'he would, wouldn't he' was said of Lord Astor by Mandy <u>Rice</u>-Davies (b) prandial nudity is a Naked Lunch, a novel by William Seward <u>Burroughs</u> (c). A partial confusion of the names leads to Edgar Rice Burroughs (d) whose most famous creation, Tarzan (e) was played in films by Johnny Weissmuller (f) who set the 100 metres freestyle record in 1922.

165.

Bizarre Pokémon is an anagram of Man Booker Prize.

(JB,2005) = The Sea by John Banville (2005 winner)
(IM,1978) = The Sea, The Sea by Iris Murdoch (1978 winner)

(a) **(JB,1972)**   = **G** by John Berger (1972 winner)
(b) **(DBCP,2003) = Vernon God Little** by D. B. C. Pierre (2003 winner)

166.

(**Maltese**) falcon, (**Portuguese**) man o' war, (**Swedish**) chef, (**Italian**) job, (**Dutch**) courage, (**Spanish**) fly, (**French**) connection, (**German**) shepherd, (**Irish**) stew, (**Norwegian**) wood, (**Turkish**) delight

167.

Villages in England: Bicker Gauntlet (Lincolnshire), Curry Mallet (Somerset), Martyr Worthy (Hampshire), Queen Camel (Somerset), Stalling Busk (North Yorkshire)

168.

Film titles containing numbers. The number is used to determine the point size of the font for the remaining letters:

12 ANGRY MEN	APOLLO 13
16 CANDLES	CATCH 22
27 DRESSES	JENNIFER 8
21 GRAMS	OCEAN'S 11
10 LITTLE INDIANS	THE MAGNIFICENT 7
9 MONTHS	TRACK 29

169.

170.

EDMONTON is missing. They are the province/territory capitals of Canada.

C	C	A	N	A	D	G	E	P	I	N	N	I	W	A	C
A	H	N	A	D	A	C	A	N	S	A	D	A	H	C	A
N	A	A	D	A	C	H	A	N	A	T	D	A	I	C	A
N	A	D	R	A	A	C	A	Y	N	A	J	D	T	A	C
A	N	A	D	L	A	C	A	T	N	T	N	O	E	A	D
A	C	A	I	R	O	T	C	I	V	O	A	N	H	A	D
A	C	F	A	N	A	T	D	C	T	R	A	C	O	N	A
N	A	A	D	A	C	A	T	C	N	O	A	D	R	A	S
X	C	A	N	A	D	A	I	E	C	N	A	N	S	A	D
A	C	A	N	A	D	R	R	B	T	T	A	C	E	A	N
A	D	A	C	A	E	N	A	E	I	O	D	A	C	A	N
A	D	A	C	D	A	N	A	U	G	D	W	A	C	A	N
A	D	A	E	C	A	N	L	Q	A	I	D	N	A	C	A
N	A	R	D	A	C	A	A	N	A	D	N	A	C	A	N
A	F	D	A	C	Q	A	N	A	D	A	C	A	A	N	A
D	A	E	F	I	N	K	W	O	L	L	E	Y	C	A	N

171.

(a) **Salisbury** becomes **Harare**
Constantinople becomes **Istanbul**
Batavia becomes **Jakarta**
Leopoldville becomes **Kinshasa**
Bombay becomes **Mumbai**
Kristiania becomes **Oslo**
Edo becomes **Tokyo**

(b) **New France** becomes **Canada**
French Somaliland becomes **Djibouti**
Abyssinia becomes **Ethiopia**
Concepcion becomes **Grenada**

Basutoland becomes **Lesotho**
Aden becomes **Yemen**
Rhodesia becomes **Zimbabwe**
Upper Volta becomes **Burkina Faso**

172.

(a) **P.** ARIES, TAURUS, ..., AQUARIUS, PISCES

(b) **E.** JANUARY, FEBRUARY, ..., NOVEMBER, DECEMBER

(c) **N.** ONE, TWO, ..., NINE, TEN

(d) **T.** MERCURY, VENUS, ..., URANUS, NEPTUNE

(e) **O.** ATHENS, PARIS, ..., HELSINKI, MELBOURNE (Summer Olympic cities)

173.

Top 20 hits.

(a) GOLDEN **BROWN** SUGAR

(b) TRUE **BLUE MOON** RIVER

(c) WONDERFUL **DREAM BABY LOVE** MACHINE

(d) BLACKBERRY **WAY DOWN DOWN UNDER** PRESSURE

(e) MA **BAKER STREET DANCE ON FIRE** BRIGADE

174.

(a) **STYLE.** E at positions 1, 2, 3, 4, 5. Rest are consonants.

(b) **LARK.** Each bird starts with the second letter of the previous bird

(c) **SHRED.** In the words A, B, C, etc. can change to B, C, D, etc. SHRED → SIRED.

(d) **KILL.** Words contain AB, BC, CD, etc.

(e) **OPAQUE.** Words contain JKL, KLM etc. OPaQue.

175.

(a) Musical note lengths in decreasing order: Breve, Semibreve, Minim, Crotchet, Quaver, Semiquaver, Demisemiquaver, Hemidemisemiquaver = **DSQ, HDSQ**

(b) Runs of Oxford and Cambridge boat race wins, 1954–1985: O^{10}

(c) Plimsoll line marks: **WNA**

176.

Zoltan (or any male name beginning with Z). The initial letters spell WALTZ, TANGO, RUMBA. (Italicised/highlighted names are not mentioned in the question.)

Professional	Celebrity	Celebrity Job
William	Teresa	Reporter
Anna	*Anthony*	*Unicyclist*
Lilia	Nigel	Model
Teresa	Greg	Businessman
Zoltan	*Olivia*	*Actress*

177.

Some Like It Hot: HANDsome, LONEsome, TWOsome, WHOLEsome; LADYlike, LIFElike, SUCHlike, WORKMANlike; BANDit, DIGit, ORBit, PULPit; EARShot, GUNShot, MUGShot, SNAPShot.

178.

In Morse code:

OM = ---/-- = ----- = 0

JT = .---/- = .---- = 1

UM = ..-/-- = ..--- = 2

VT = ...-/- = ...-- = 3

SA = .../.- = ...- = 4

IS = ../... = ... = 5

BE = -.../. = -.... = 6

TB = -/-... = --... = 7

OI = ---/.. = ---.. = 8

MG = --/--. = ----. = 9

179.

(a) Number of letters (4/5/6/7/8) in NATO phonetic name of the letters (e.g. E=ECHO/4, A=ALPHA/5, ..., N=NOVEMBER/8)

(b) Number of characters (1/2/3/4) in Morse code representation of the letters (e.g. E=./1, A=.-/2, D=-../3, B=-.../4)

(c) Number of raised 'dots' (1/2/3/4/5) in the Braille representation of the letter

(d) Prime/non-prime numbers numerically (A=1, B=2, ..., Z=26)

180.

CHANCE (Arm) fits between ACID (Ant) and HARD (Ball).

Ant	**Arm**	Ball	Bat		Bliss	Break
ACID	**CHANCE**	HARD	HER		WEDDED	CLEAN
Brussels		Can	Cow		Cross	Cue
SPROUT		ADA	BOY		GEORGE	ON
Cushion		Den	Dog		Dress	Eleven
PIN		MARK	GONE		PARTY	OCEAN'S
Emu		Face	Feet		Forty	Forty-Two
LATE		SHOW	FIND		TOP	LEVEL
Foul		Fox	Germ		Grass	Hand
PROFESSIONAL		GLOVE	ANY		PAMPAS	GIVE
Head		Heart	Hen		Hun	Ice
USE		CROSS	BANE		GARY	LAND
Jay		Leg	Lima		London	Manila
WALK		PULL	BEAN		PRIDE	ENVELOPE
Miss		Ness	Nether		Ninety	Ninety-Five
NEAR		LOCH	LANDS		JOE	WINDOWS
Nor		Pass	Pocket		Pot	Prague
WAY		FORWARD	PICK		FLOWER	SPRING
Sixty-Six		Spa	Stockholm		Table	Ten
ROUTE		IN	SYNDROME		TIMES	NUMBER
Toes		Toss	Twenty-Two		Vienna	Wellington
TOUCH		CABER	CATCH		WOODS	BOOT

Themes are:

Can be prefixed by a 3-letter creature: *ANT*ACID, *BAT*HER, *COW*BOY, *DOG*GONE, *EMU*LATE, *FOX*GLOVE, *HEN*BANE, *JAY*WALK

X your Y (Y is a part of the body): **CHANCE** your ***ARM***, SHOW your *FACE*, FIND your *FEET*, GIVE your *HAND*, USE your *HEAD*, CROSS your *HEART*, PULL your *LEG*, TOUCH your *TOES*

Snooker/billiards/pool term: HARD *BALL*, CLEAN *BREAK*, ON *CUE*, PIN *CUSHION*, PROFESSIONAL *FOUL*, PICK*POCKET*, FLOWER*POT*, TIMES *TABLE*

Can be followed by a 4/5-letter word ending -SS: WEDDED *BLISS*, GEORGE *CROSS*, PARTY *DRESS*, PAMPAS *GRASS*, NEAR *MISS*, LOCH *NESS*, FORWARD *PASS*, CABER *TOSS*

Can be preceded by a capital city: *BRUSSELS* SPROUT, *LIMA* BEAN, *LONDON* PRIDE, *MANILA* ENVELOPE, *PRAGUE* SPRING, *STOCKHOLM* SYNDROME, *VIENNA* WOODS, *WELLINGTON* BOOT

Can be prefixed by a word to form a NATO nation: *CAN*ADA, *DEN*MARK, *GERM*ANY, *HUNG*ARY, *ICE*LAND, *NETHER*LANDS, *NOR*WAY, *SP*AIN

Can be followed by a 2-digit number: OCEAN'S *ELEVEN*, TOP *FORTY*, LEVEL *FORTY-TWO*, JOE *NINETY*, WINDOWS *NINETY-FIVE*, ROUTE *SIXTY-SIX*, NUMBER *TEN*, CATCH *TWENTY-TWO*

181.

First words of consecutive books of the New Testament (King James version), starting at St John's Gospel. Next is the book of Hebrews, beginning with the word **God**.

182.

(a) **Characters in The Hobbit**: Kili, Dori, Fili, Bilbo, Balin, Nori (Bilbo is an alternative spelling of Bilbao)

(b) **Repeated strings**: Ndanda, Kankan, Tin Tin, Xai-Xai, Wagga Wagga, Walla Walla

183.

AGE.

1: anagram of ONE, EON, a synonym of AGE

2 tutus: ARCHBISHOP, SKIRT

3 members of sets of 3: IRINA (Sisters), LARRY (Stooges), READING (Rs)

4 Calling Birds (i.e. homophones of birds): CHUFF/chough, RENNES/wren, SKEWER/skua, TURN/tern

Famous Five/5 (synonyms of 'famous'): KNOWN, NOTED, NOTORIOUS, RECOGNISED, RENOWNED

Words (of lengths 6–11 letters) with F (the 6th letter of the alphabet) as their 6th letter: BELIE$\underline{F}$, GRAPE$\underline{F}$RUIT, MAGNI$\underline{F}$ICENT, MIDWI$\underline{F}$E, QUALI$\underline{F}$IED, TRANS$\underline{F}$ER

Seven/7 seas: BATTERsea, BLACK sea, CORAL sea, DEAD sea, IRISH sea, NORTH sea, RED sea

8-letter words containing all the letters of EIGHT: C$\underline{HEATING}$, GL$\underline{ITCHES}$, $\underline{G}$RAP$\underline{HITE}$, $\underline{HI}$STO$\underline{GE}$N, K$\underline{NIGHTED}$, M$\underline{EGALITH}$, OUT$\underline{WEIGH}$, $\underline{THE}$UR$\underline{GIC}$

Nine/9 pins: DRAWING pin, FIRING pin, HAIRpin, HATpin, KINGpin, ROLLING pin, SAFETY pin, TAILSpin, TIEpin

Can be preceded/followed by TEN/10: tenABLE, tenANT, CHRISten, tenDON, tenDRILLED, FLATten, FORGOTten, HASten, tenOR, ROTten

184.

(a) **113 = 7^2 + 2^6**: 2 = 1^2 + 2^0, 6 = 2^2 + 2^1, ..., 68 = 6^2 + 2^5 (squares + powers of 2)

(b) **57 = 23 + 34**: 3 = 2 + 1, 4 = 3 + 1, ..., 40 = 19 + 21 (prime numbers + Fibonacci series)

(c) **7 = 5 + 2**: 5 = 3 + 2, 8 = 1 + 7, ..., 14 = 6 + 8 (digits of e = 2.71828182... + digits of π = 3.14159265...)

(d) **17 = 11 + 6**: 4 = 1 + 3, ..., 13 = 10 + 3 (numbers + length of number in English)

(e) **14 = 6 + 8**: 6 = 2 + 4, 8 = 4 + 4, ..., 10 = 5 + 5 (length of number in French + length of number in German)

(f) **29 = 17 + 12**: 4 = 1 + 3, 8 = 2 + 6, ..., 30 = 16 + 14 (number + Scrabble score of number in English)

185.

SHREW. All references are to the title of a Shakespeare play: The Merry Wives of WINDSOR (a town in southern England); The Merchant of VENICE and Two Gentlemen of VERONA (two cities in Italy); Timon of ATHENS (a city in Greece); Pericles, Prince of TYRE (a city in Lebanon) and The Taming of the SHREW.

186.

Viva La Vida.

O		H	A	P	P	Y	A	S	A	S	A	N	D	B	O	Y
F		A		R		L		W		P		I		V		
F	L	Y	S	P	E	C	K		R	H	E	O	S	T	A	T
T		E		M		A		Y		X		O				A
H	A	S	B	R	O	S	L	D	A	I	E	V	R			T
E			U	E	L	V	I	R	O	S	A	D	D	E	S	T
B			G	D	A	I	E	S	V	L	R	O	E			E
E	Y	E	S	O	R	E	S	L	D	V	I	A	R			R
A			L	V	D	A	I	E	R	O	S	E	B	U	D
T	O	P	H	I	S	A	O	V	R	E	D	L	D			E
E		U		V	I	R	D	A	S	O	L	E		E		M
N	O	T	E	S	D	O	R	E	L	A	V	I	N	D	I	A
T		O		A	E	L	V	O	I	D	S	R		U		L
R	I	N	G	O	U	S	E	L		U		E	X	C	E	L
A		I		S		I		I	B	S	E	N		A		I
C		C		L		D		A		E		I		T		O
K	E	E	P	O	N	E	S	N	O	S	E	C	L	E	A	N

187.

(a) **FLYWEIGHT** is the only 2-syllable word among 4-syllable words.

(b) **HOSPITABLE** is the only word not to contain all 5 vowels.

(c) **CENTRIFUGAL** is the only word not to contain 5 consecutive letters of the alphabet.

188.

PNEUMONOULTRAMICROSCOPICSILICOVOLCANOCONIOSIS.
Place the groups into a 9×5 grid and read up the columns from right to left.

```
S  N  L  O  U
I  A  I  R  O
S  C  S  C  N
O  L  C  I  O
I  O  I  M  M
N  V  P  A  U
O  O  O  R  E
C  C  C  T  N
O  I  S  L  P
```

189.

ARKANSAS. The words contain another word in the same category: ASTATINE (elements), ELEPHANT (creatures), EPSILON (Greek letters), ROMANIA (countries), SUSANNAH (Girls' names), WALTER (Boys' names), WEST VIRGINIA and ARKANSAS (US states).

190.

(a) **19.** Convert the sequence to letters and it spells out CHRISTMAS.

(b) **10.** These are in order around a dartboard.

191.

'Monday's Child':

Is Fair Of Face – Is Full Of Grace – Is Full Of Woe + Is Loving And Giving + Works Hard For A Living

= Monday – Tuesday – Wednesday + Friday + Saturday

= Sunday = Is Bonny And Blithe And Good And Gay =
IBABAGAG

Addition and subtraction is done by assigning the numbers 1 to 7 to the days Monday to Sunday (i.e. using the line numbers of the poem): 1-2-3+5+6=7.

192.

Film directors and their works:

Allen (Woody)	= Everything You Always Wanted To Know About Sex *But Were Afraid To Ask, Annie Hall, Manhattan, Hannah And Her Sisters
Burton (Tim)	= Beetle Juice, Edward Scissorhands, Ed Wood, Planet Of The Apes
Cameron (James)	= The Terminator, Aliens, True Lies, Titanic
De Palma (Brian)	= Carrie, The Untouchables, Bonfire Of The Vanities, Mission:Impossible
Eastwood (Clint)	= Play Misty For Me, Unforgiven, The Bridges Of Madison County, Mystic River
Ford (John)	= The Grapes Of Wrath, How Green Was My Valley, The Searchers, The Man Who Shot Liberty Valance
Gilliam (Terry)	= Monty Python And The Holy Grail, Time Bandits, Brazil, Twelve Monkeys
Hitchcock (Alfred)	= The Lady Vanishes, The Man Who Knew Too Much, North By Northwest, Psycho
Ivory (James)	= Heat And Dust, A Room With A View, Remains Of The Day, Howard's End
Jackson (Peter)	= Bad Taste, Heavenly Creatures, Lord Of The Rings, King Kong

Kubrick (Stanley) = Paths Of Glory, Doctor Strangelove Or: How I Learnt To Stop Worrying And Love The Bomb, A Clockwork Orange, The Shining

Landis (John) = Kentucky Fried Movie, The Blues Brothers, An American Werewolf In London, Coming To America

Mann (Michael) = Manhunter, The Last Of The Mohicans, Heat, Miami Vice

Nolan (Christopher) = Memento, Insomnia, Batman Begins, The Prestige

Oz (Frank) = The Dark Crystal, The Muppets Take Manhattan, Little Shop Of Horrors, The Stepford Wives

Polanski (Roman) = Cul-De-Sac, Rosemary's Baby, Chinatown, The Pianist

Quine (Richard) = Bell Book And Candle, The World Of Susie Wong, Sex And The Single Girl, How To Murder Your Wife

Reiner (Rob) = This Is Spinal Tap, Stand By Me, When Harry Met Sally, Misery

Spielberg (Steven) = Jaws, Raiders Of The Lost Ark, Jurassic Park, Schindler's List

Tarantino (Quentin) = Reservoir Dogs, Pulp Fiction, Jackie Brown, Death Proof

Underwood (Ron) = Tremors, City Slickers, Mighty Joe Young, The Adventures Of Pluto Nash

Verhoeven (Paul) = Total Recall, Basic Instinct, Starship Troopers, Hollow Man

Wilder (Billy) = The Seven Year Itch, Witness For The Prosecution, Some Like It Hot, The Private Life Of Sherlock Holmes

Xie (Jin) = Gao Shan Xia De Hua Huan, Zui Hou De Gui Zu, Qing Liang Si Zhong Sheng, Bai Lu Yuan

Young (Terence) = Dr No, From Russia With Love, Thunderball, Wait Until Dark

Zemeckis (Robert) = Back To The Future, Who Framed Roger Rabbit, Forrest Gump, Cast Away

193.

Five. The completed square is:

```
U  N  D  E  U  X
T  R  O  I  S  Q
U  A  T  R  E  C
I  N  Q  S  I  X
S  E  P  T  H  U
I  T  N  E  U  F
```

194.

KEY-PHRASE. The sentence given is 'TO BE OR NOT TO BE THAT IS THE KEY-PHRASE' enciphered via:

ABCDEFGHIJKLMNOPQRSTUVWXYZ
TOBEORNOTTOBETOBEORNOTTOBE

195.

(a) **Raging Bull**   (f) **Con Air**

(b) **Psycho**   (g) **Heat**

(c) **My Fair Lady**   (h) **Ben Hur**

(d) **Mad Max**   (i) **Alien**

(e) **Die Hard**   (j) **Flaming Star**

196.

(a) **OILED** has its letters in neither lexicographic (BEGIN, ADEPT, BELOW, FIRST, GIPSY, FORTY) nor antilexicographic (POLKA, TRIED, SOLID, TONIC, SPOKE, WRONG) order.

(b) **SIEGE** contains letters from both the top row (EERIE, TEPEE, PIPER, QUITE, WRITE) and bottom 2 rows (CLASH, CHASM, NAVAL, SHANK, GLASS) of a typewriter.

(c) **ORGAN** contains neither 0 (BEING, VAPID, WRECK, FOCUS, EXTRA, SQUID) nor 4 (FACED, CUBED, FIGHT, LEMON, BURST, TORUS) consecutive letters of the alphabet.

(d) **GAMES** is not an anagram of either a boy's (CLEAN, CIGAR, BAILS, WINED, SMILE, LYRIC, WILES) or a girl's (LADLE, HOARD, DAILY, AURAL, CANNY, MANOR, BYTES) name.

(e) **MERIT** is not a word backwards (LACED, MINED, SPEED, DEPOT, LEVER, REBUT) or cyclically shifted (ANGLE, LEASE, ZEBRA, ELBOW, ALLOY, VERSE).

In each part the words are in order of their letter sum (A=1, B=2, ..., Z=26).

197.

AFFIXES (1 across) and **ASIDE** (1 down).

A 1	F 2	F 2	I 3	X 4	E 5	S 6	■	A 1	B 7	I 3	D 8	E 5
S 6	■	I 3	■	■	■	W 9	■	P 10	■	T 11	■	U 12
I 3	O 13	T 11	A 1	■	J 14	E 5	R 15	R 15	Y 16	C 17	A 1	N 18
D 8	■	F 2	■	A 1	■	L 19	■	O 13	■	H 20	■	U 12
E 5	Q 21	U 12	I 3	V 22	A 1	L 19	E 5	N 18	T 11	■	■	C 17
■	■	L 19	■	O 13	■	S 6	■	■	■	C 17	■	H 20
A 1	L 19	L 19	O 13	W 9	S 6	■	T 11	H 20	R 15	O 13	W 9	S 6
N 18	■	Y 16	■	■	F 2	■	U 12	■	A 1	■	■	■
A 1	■	■	C 17	I 3	T 11	R 15	O 13	N 18	E 5	L 19	L 19	A 1
E 5	■	A 1	■	N 18	■	O 13	■	G 23	■	F 2	■	D 8
M 24	O 13	L 19	L 19	U 12	S 6	C 17	S 6	■	H 20	A 1	L 19	O 13
I 3	■	S 6	■	R 15	■	K 25	■	■	■	C 17	■	B 7
A 1	B 7	O 13	D 8	E 5	■	S 6	Q 21	U 12	E 5	E 5	Z 26	E 5

01	02	03	04	05	06	07	08	09	10	11	12	13	14	15	16	17	18	19	20	21	22	23	24	25	26
A	F	I	X	E	S	B	D	W	P	T	U	O	J	R	Y	C	N	L	H	Q	V	G	M	K	Z

198.

    (a) **SEASONAL GREETINGS TO YOU ALL**

    (b) **HARK THE HERALD ANGELS SING GL**(ory to the newborn king)

Letter-by-letter and with A=1, B=2, ..., Z=26

```
AFSDIVFTLJFQXJBNXFHRXIHSX =
HARKTHEHERALDANGELSSINGGL +
SEASONALGREETINGSTOYOUALL
QXSTGVFXZBXILTBBNTFVVUHWB =
HARKTHEHERALDANGELSSINGGL -
SEASONALGREETINGSTOYOUALL
```

199.

    (a) **SIX**.

    (b) **VI** was hidden instead.

> *Dear Julius,*
>
> *I have d<u>one</u> what you suggested and <u>it w</u>orked! We both <u>re-engineered</u> and reconstructed the thing, and all o<u>f our</u> efforts were rewarded. But, just to see i<u>f I've</u> been a fool and missed something <u>vi</u>tal, we spent thi<u>s eve</u>ning taking the sl<u>eigh</u> to bits, and rebuilding it, yet agai<u>n</u>. <u>I</u> n<u>eve</u>r thought it would take so long, but the nigh<u>t en</u>ded happily. Hoorah!*
>
> *With much gratitude*
>
> *Santa Claus*

200.

    (a) **BINDER**. Made up of two German words, rather than two French words.

    (b) **TRACTORS**. Contains anagram of vegetable in middle, rather than fruit.

    (c) **HUNTERS**. Last letter can move to front to form new word, rather than first letter move to end.

(d) **NECROMANCY**. Start sounds like part of the body, rather than end.

(e) **ALLIANCE**. Even letters form word, rather than odd letters.

201.

FORTUNE (Outrageous) fits between EMPIRE (Ottoman) and DOC (Papa).

A	And	Armchair		Be	Bean
SEA	ARROWS	THEATRE		THAT	BAG
Bed	Boys	Cabbage		Cabinet	California
SIT	BEACH	WHITE		PUDDING	GIRLS
Cauliflower	Chair	Charlie		Chocolate	Couch
EAR	LIFT	BROWN		HOT	POTATO
Fail	Gees	Golf		Hail	Hotel
SAFE	BEE	COURSE		STONE	CALIFORNIA
India	Indiana	Jail		Kentucky	Kilo
RUBBER	JONES	BIRD		FRIED	GRAM
Maiden	Mail	Marrow		Minnesota	Mushroom
IRON	BOX	BONE		FATS	CLOUD
Nail	Onion	Ottoman	**Outrageous**	Papa	Pea
BITING	RINGS	EMPIRE	**FORTUNE**	DOC	SHOOTER
Pepper	Pistols	Question		Quo	Rail
MINT	SEX	WHETHER		STATUS	WAY
Rhode Island	Sail	Sierra		Stones	Stool
RED	OR	NEVADA		ROLLING	PIGEON
Straits	Table	Tail		Take	Tennessee
DIRE	FOOTBALL	BACK		ARMS	WILLIAMS
The	To	Virginia		Washington	Yankee
MIND	SUFFER	CREEPER		IRVING	DOODLE

Themes are:

To be or not to be, preceding words: A SEA, And ARROWS, Be THAT, **Outrageous FORTUNE,** Question WHETHER, Take ARMS, The MIND, To SUFFER

Furniture: Armchair THEATRE, Bed SIT, Cabinet PUDDING, Chair LIFT, Couch POTATO, Ottoman EMPIRE, Stool PIGEON, Table FOOTBALL

Vegetables: Bean BAG, Cabbage WHITE, Cauliflower EAR, Marrow BONE, Mushroom CLOUD, Onion RINGS, Pea SHOOTER, Pepper MINT

BANDS: BEACH Boys, HOT Chocolate, BEE Gees, IRON Maiden, SEX Pistols, STATUS Quo, ROLLING Stones, DIRE Straits

US states: California GIRLS, Indiana JONES, Kentucky FRIED, Minnesota FATS, Rhode Island RED, Tennessee WILLIAMS, Virginia CREEPER, Washington IRVING

Nato phonetics: Charlie BROWN, Golf COURSE, Hotel CALIFORNIA, India RUBBER, KiloGRAM, Papa DOC, Sierra NEVADA, Yankee DOODLE

–AIL: FailSAFE, HailSTONE, JailBIRD, MailBOX, NailBITING, RailWAY, SailOR, TailBACK

202.

My daughter has a set of alphabet pictures, laid out in 4 rows of 7. These sentences are based on pictures going down the columns. I had Apple, Hat, Octopus and Volcano; my daughter had Bee, Igloo, Penguin and Wigwam but she misidentified the Igloo as a Cave. (Based on a true story!).

203.

Paroxysm: taking alternate letters yields the whole alphabet without repeats.

BEDAZZLEMENTHIGHJACKQUANTIFYUNVEILEDYOKEWOODPAROXYSM
B.D.Z.L.M.N.H.G.J.C.Q.A.T.F.U.V.I.E.Y.K.W.O.P.R.X.S.

204.

Each state has the property of having one example of the nth letter of the alphabet as its nth letter:

ALABAMA, ALASKA, ARIZONA, ARKANSAS, CALIFORNIA, CONNECTICUT, MAINE, MICHIGAN, MINNESOTA, NEW JERSEY, NEW MEXICO, TENNESSEE, WASHINGTON, WEST VIRGINIA, WYOMING

RHODE ISLAND has two examples.

205.

WICKLOW. The set is that of the 26 counties of the Republic of Ireland. If sorted in alphabetic order and each letter of the nth in the list is advanced by n letters (cyclically) in the alphabet, the entries of the table are generated. WICKLOW, the last in the list, remains unchanged when its letters are advanced by 26 characters. In alphabetic order the 'enciphered' entries are:

LOUTH+15=ADJIW, MAYO+16=CQOE, CARLOW+1=DBSMPX, MEATH+17=DVRKY, CAVAN+2=ECXCP, MONAGHAN+18=EGFSYZSF, CLARE+3=FODUH, CORK+4=GSVO, OFFALY+19=HYYTER, DONEGAL+5=ITSJLFQ, DUBLIN+6=JAHROT, ROSCOMMON+20=LIMWIGGIH, SLIGO+21=NGDBJ, GALWAY+7=NHSDHF, TIPPERARY+22=PELLANWNU, KERRY+8=SMZZG, KILDARE+9=TRUMJAN, WATERFORD+23=TXQBOCLOA, WESTMEATH+24=UCQRKCYRF, KILKENNY+10=USVUOXXI, WEXFORD+25=VDWENQC, WICKLOW+26=WICKLOW, LAOIS+11=WLZTD, LEITRIM+12=XQUFDUY, LIMERICK+13=YVZREVPX, LONGFORD+14=ZCBUTCFR

206.

(a) (I CAN'T GET NO) SATISFACTION; PAINT IT, BLACK (The Rolling Stones)

(b) PLEASE PLEASE ME; HEY JUDE (The Beatles)

(c) THE SURREY WITH THE FRINGE ON TOP; THE FARMER AND THE COWMAN (Oklahoma!)

(d) YOU KNOW MY NAME; ANOTHER WAY TO DIE (James Bond Themes)

The encryption is via simple substitution. In (a), for example, 'PAINT IT, BLACK' is encrypted via:

```
ABCDEFGHIJKLMNOPQRSTUVWXYZ
ICANTGEOSFBDHJKLMPQRUVWXYZ
```

and the encryption of '(I CAN'T GET NO) SATISFACTION' is via:

```
ABCDEFGHIJKLMNOPQRSTUVWXYZ
PAINTBLCKDEFGHJMOQRSUVWXYZ
```

207.

'Is it some education' is an anagram of Situation Comedies.

(a) **To the Manor Born**

(b) **Are You Being Served?**

(c) **Last of the Summer Wine**

(d) **Blackadder Goes Forth**

(e) **The Vicar of Dibley**

208.

(a) **ACADEMICAL.** Words of increasing length formed alternately from A-M, N-Z.

(b) **AMBIDEXTROUS.** Half the word from A-M, half from N-Z, alternating, word length increasing by 2.

(c) **JONQUIL.** Letters 1 and 4 move through the alphabet. Lengths alternate 6/7.

(d) **MAZE.** Words contain A, B, C, etc. and N, O, P, etc. MAZE contains M and Z.

209.

Atomic numbers and their chemical symbols: W H At, I S, O Ne, P Lu S, Ni Ne? **Te N =52 7**

210.

CAP.

1 (One) can be added: CAPone

2 Turtle Doves: TAUT, HORS D'OEUVRES

3 Graces: grace DARLING, grace JONES, grace KELLY

4 quarters: quarter-BACK, quarter-DECK, quarter-FINAL, quarterMASTER

The Jackson 5: jackson BROWNE, COLIN jackson, GLENDA jackson, PERCY jackson, jackson POLLOCK

Can be preceded by 6/VI: viCAR, viEWER, viPERISH, viRILE, viSAGE, viTALLY

7 ages: BRONZE age, GOLDEN age, IRON age, MIDDLE age, OLD age, SPACE age, STONE age

Synonyms rhyme with 8/eight: DESTINY/fate, DETEST/hate, PAL/mate, PORTAL/gate, SATISFY/sate, TARDY/late, TRYST/date, VALUE/rate

9 9-letter words, first letters spell SEPTEMBER, last letters spell MONTH NINE:

S	T	R	O	N	T	I	U	M
E	S	P	E	R	A	N	T	O
P	R	O	T	O	Z	O	O	N
T	R	A	N	S	P	O	R	T
E	S	T	A	B	L	I	S	H
M	A	G	N	E	T	R	O	N
B	E	R	N	O	U	L	L	I
E	D	U	C	A	T	I	O	N
R	E	C	T	A	N	G	L	E

10 Green Bottles: LINCOLN green, PUTTING green, REVEREND green, SEA-green, VILLAGE green, BLUEbottle, KLEIN bottle, MILK bottle, WATER bottle, WINE bottle

211.

In each case the name of the theme (the 'precise answer') is the encipherment key.

(a) **COLOURS OF THE RAINBOW**: ORANGE, RED, INDIGO, BLUE, GREEN, VIOLET, YELLOW

```
COLURSFTHEAINBWDGJKMPQVXYZ
ABCDEFGHIJKLMNOPQRSTUVWXYZ
```

(b) **SNOW WHITE'S SEVEN DWARFS**: SNEEZY, SLEEPY, HAPPY, DOC, DOPEY, BASHFUL, GRUMPY

```
SNOWHITEVDARFBCGJKLMPQUXYZ
ABCDEFGHIJKLMNOPQRSTUVWXYZ
```

(c) **THE SEVEN AGAINST THEBES**: TYDEUS, HIPPOMEDON, ADRASTUS, AMPHIARAUS, CAPANEUS, PARTHENOPEUS, POLYNICES

```
THESVNAGIBCDFJKLMOPQRUWXYZ
ABCDEFGHIJKLMNOPQRSTUVWXYZ
```

212.

(a) Pop stars: (Dollar) – (50 Cent) = 50 Cent = **Curtis Jackson**

(b) Actors in The Prisoner: Number 2 + Number 2 + Number 2 = Number 6 = **McGoohan**

(c) Juror numbers in the film Twelve Angry Men: $(2 \times 9) - 11 = 7 =$ **Warden**

(d) UK chart singles: $((1999 + 19 + 17 - 54) \times 1/10) - 98.6 = 99\frac{1}{2} =$ **Carol Lynn Townes**

213.

(a) **Top Gun**

(b) **Up**

(c) **Star Wars**

(d) **Mona Lisa**

(e) **Easy Rider**

(f) **Dr No**

(g) **Annie Hall**

(h) **La Haine**

(i) **El Cid**

(j) **Turk 182**

214.

The themes are:

(a) **Grammatical cases**: ABLATIVE, ACCUSATIVE, DATIVE, GENITIVE, NOMINATIVE, VOCATIVE

(b) **Chess pieces**: BISHOP, CASTLE, KING, KNIGHT, PAWN, QUEEN

(c) **States of New England**: CONNECTICUT, MAINE, MASSACHUSETTS, NEW HAMPSHIRE, RHODE ISLAND, VERMONT

(d) **Noble gases**: ARGON, HELIUM, KRYPTON, NEON, RADON, XENON

(e) **Counties of Northern Ireland**: ANTRIM, ARMAGH, DOWN, FERMANAGH, LONDONDERRY, TYRONE

(f) **Wives of Henry VIII**: ANNE, ANNE, CATHERINE, CATHERINE, CATHERINE, JANE

1	1	6	6	5	4
ABLATIVE	ACCUSATIVE	ANNE	ANNE	ANTRIM	ARGON
5	2	2	6	6	6
ARMAGH	BISHOP	CASTLE	CATHERINE	CATHERINE	CATHERINE
3	1	5	5	1	4
CONNECTICUT	DATIVE	DOWN	FERMANAGH	GENITIVE	HELIUM
6	2	2	4	5	3
JANE	KING	KNIGHT	KRYPTON	LONDONDERRY	MAINE
3	4	3	1	2	2
MASSACHUSETTS	NEON	NEW HAMPSHIRE	NOMINATIVE	PAWN	QUEEN
4	3	5	3	1	4
RADON	RHODE ISLAND	TYRONE	VERMONT	VOCATIVE	XENON

The KING and QUEEN (chess pieces) are associated with the Queens of King Henry VIII.

215.

Our best answer is

BLACKSMITH and **GUNPOWDER**: 10 × 9 = 90

216.

(a) **The elements with single letter symbols**

(b) **CDILMVX** (Roman numerals), **AJKQ** (picture playing cards), **ENSW** (points of the compass) and **FP** (hard and soft, musically)

217.

(a) **IMAGE**. In all the others letters 1, 3 and 5 form a word: AGE, BIG, CAP, DRY, EAT, FED, GAS, HAT, JUT

(a) **DRINK**. In all the others letters 2 and 4 form a word: BE, US, OR, NO, ON, IN, AM, DO, UP

218.

	Harrison	*Johnson*	*Jackson*	*Washington*
Singers	George	Holly	Michael	Geno
US Presidents	Benjamin	Andrew	Andrew	George
Actors	Rex	Don	Gordon	**Denzel**

219.

(a) They are the only digraphs that are both US state abbreviations and symbols of elements.

(b) **WISCONSIN**. There are only 6 US states that can be spelled using the symbols of the elements and WISCONSIN is the only one that can be thus spelled in more than one way.

MONTANA (1): MO-N-TA-NA

NEWHAMPSHIRE (1): NE-W-H-AM-P-S-H-I-RE

OHIO (1): O-H-I-O

SOUTHCAROLINA (1): S-O-U-TH-C-AR-O-LI-NA

UTAH (1): U-TA-H

WISCONSIN (9): W-I-S-C-O-N-SI-N, W-I-S-C-O-N-S-IN, W-I-S-C-O-N-S-I-N, W-I-S-CO-N-SI-N, W-I-S-CO-N-S-IN, W-I-S-CO-N-S-I-N, W-I-SC-O-N-SI-N, W-I-SC-O-N-S-IN, W-I-SC-O-N-S-I-N

220.

QUOTE.

Eliminating those words whose 1ˢᵗ letter is W (WASTE, WEAVE, WHIRL, WORTH, WRONG) leaves:

ALGAE, BULBS, COUCH, DEPTH, EASEL, FJORD, GENII, HOPED, JAMMY, KNIFE, LABEL, MAIMS, NIECE, ORGAN, PATHS, QUOTE, SEATS, TITHE, UPSET, VERVE, YACHT

Eliminating those words whose 2ⁿᵈ letter is a consonant (ALGAE, FJORD, KNIFE, ORGAN, UPSET) leaves:

BULBS, COUCH, DEPTH, EASEL, GENII, HOPED, JAMMY, LABEL, MAIMS, NIECE, PATHS, QUOTE, SEATS, TITHE, VERVE, YACHT

Eliminating those words whose 3rd letter scores 3 in Scrabble (DEPTH, HOPED, JAMMY, LABEL, YACHT) leaves:

BULBS, COUCH, EASEL, GENII, MAIMS, NIECE, PATHS, QUOTE, SEATS, TITHE, VERVE

Eliminating those words whose 4th letter is the same as the 1st (BULBS, COUCH, EASEL, MAIMS, VERVE) leaves:

GENII, NIECE, PATHS, QUOTE, SEATS, TITHE

Eliminating those words whose 5th letter is the sum (mod 26 with A=1, B=2, ..., Z=26) of the 1st–4th letters (GENII, NIECE, PATHS, SEATS, TITHE) leaves:

QUOTE

221.

- (a) **4181.** These are every other Fibonacci number.
- (b) **431.** Sequence is $2^N - N^2$.
- (c) **6.** Sequence is the gaps between successive prime numbers. 23 to 29 is 6.
- (d) **7.** Sequence is decimal expansion of 1/19.
- (e) **841.** Each member of the sequence is the sum of the factors of the previous one.

222.

113323,373,373 or 113,373,373,373. Each number in the sequence is the lowest whose length is the previous number:

'Eleven' has 6 letters.

'Twenty-three' has 11 letters.

'One hundred and twenty-four' has 23 letters.

'One hundred and thirteen thousand three hundred and twenty-three million, three hundred and seventy-three thousand three hundred and seventy-three' has 124 letters.

Alternatively, for those who think 10^9 is one billion, the answer is: 'One hundred and thirteen billion, three hundred and seventy-three million, three hundred and seventy-three thousand three hundred and seventy-three', which has 124 letters.

223.

There are 47 chemical elements contained in the grid (or 48 if you are American and use the spelling Cesium instead of Caesium).

Astatine	(10,3),(8,4),(6,3),(7,1),(6,3),(7,5),(8,7),(6,6)
Barium	(4,9),(3,7),(5,6),(7,5),(6,7),(7,9)
Berkelium	(6,2),(4,3),(3,5),(4,7),(6,6),(5,4),(7,5),(6,7),(7,9)
Beryllium	(6,2),(4,3),(3,5),(1,4),(3,3),(5,4),(7,5),(6,7),(7,9)
Bohrium	(1,1),(2,3),(4,4),(3,6),(4,8),(6,7),(7,9)
Boron	(4,9),(6,8),(8,9),(6,8),(8,7)
Cerium	(1,6),(2,4),(3,6),(4,8),(6,7),(7,9)
Cesium	(8,1),(10,2),(9,4),(7,5),(6,7),(7,9)
Chlorine	(3,2),(4,4),(6,5),(7,7),(5,6),(7,5),(8,7),(6,6)
Chromium	(3,2),(4,4),(5,6),(6,8),(5,10),(4,8),(6,7),(7,9)
Curium	(1,6),(2,8),(3,6),(4,8),(6,7),(7,9)
Dubnium	(2,2),(4,1),(6,2),(8,3),(7,5),(6,7),(7,9)
Erbium	(4,3),(3,5),(2,7),(4,8),(6,7),(7,9)
Fermium	(1,9),(3,8),(1,7),(2,9),(4,8),(6,7),(7,9)
Gallium	(6,4),(5,2),(3,3),(5,4),(7,5),(6,7),(7,9)
Gold	(10,9),(8,8),(7,6),(9,5)

Hassium	(4,4),(5,2),(7,3),(9,4),(7,5),(6,7),(7,9)
Helium	(7,8),(6,6),(5,4),(7,5),(6,7),(7,9)
Iodine	(9,2),(10,4),(9,6),(7,5),(8,7),(6,6)
Iridium	(7,5),(5,6),(7,5),(9,6),(7,5),(6,7),(7,9)
Iron	(10,10),(8,9),(6,8),(8,7)
Lead	(5,4),(6,6),(7,4),(9,5)
Mercury	(5,1),(4,3),(3,5),(1,6),(2,8),(1,10),(3,9)
Neodymium	(5,7),(3,8),(2,6),(1,8),(3,9),(5,10),(4,8),(6,7),(7,9)
Nobelium	(8,7),(10,6),(8,5),(6,6),(5,4),(7,5),(6,7),(7,9)
Osmium	(6,9),(4,10),(2,9),(4,8),(6,7),(7,9)
Oxygen	(6,1),(8,2),(10,1),(9,3),(7,2),(9,1)
Polonium	(7,10),(6,8),(7,6),(6,8),(8,7),(7,5),(6,7),(7,9)
Potassium	(1,5),(2,3),(3,1),(5,2),(7,3),(9,4),(7,5),(6,7),(7,9)
Radium	(8,9),(10,8),(9,6),(7,5),(6,7),(7,9)
Radon	(8,9),(10,8),(9,6),(10,4),(8,3)
Rhenium	(8,10),(7,8),(9,9),(8,7),(7,5),(6,7),(7,9)
Rhodium	(8,10),(9,8),(7,7),(9,6),(7,5),(6,7),(7,9)
Ruthenium	(5,9),(6,7),(8,6),(7,8),(9,9),(8,7),(7,5),(6,7),(7,9)
Samarium	(5,8),(3,7),(2,5),(3,7),(5,6),(7,5),(6,7),(7,9)
Selenium	(5,8),(6,6),(5,4),(6,6),(8,7),(7,5),(6,7),(7,9)
Silver	(9,4),(7,5),(5,4),(4,6),(3,8),(5,9)
Sodium	(5,8),(7,7),(9,6),(7,5),(6,7),(7,9)
Terbium	(5,5),(4,3),(3,5),(2,7),(4,8),(6,7),(7,9)
Thallium	(6,3),(4,4),(5,2),(3,3),(5,4),(7,5),(6,7),(7,9)
Thorium	(8,6),(9,8),(7,7),(5,6),(4,8),(6,7),(7,9)
Tin	(6,3),(7,5),(8,7)
Titanium	(6,3),(7,5),(6,3),(7,1),(8,3),(7,5),(6,7),(7,9)
Uranium	(6,10),(8,9),(10,8),(8,7),(7,5),(6,7),(7,9)
Vanadium	(9,10),(10,8),(8,7),(10,8),(9,6),(7,5),(6,7),(7,9)
Xenon	(4,5),(6,6),(8,7),(6,8),(8,7)
Ytterbium	(4,2),(6,3),(5,5),(4,3),(3,5),(2,7),(4,8),(6,7),(7,9)
Yttrium	(4,2),(6,3),(5,5),(3,6),(4,8),(6,7),(7,9)

224.

SYSTEM (Truck) fits between TOUR (Tower) and BELCH (Twelfth Night).

Advice	Away	Boat		Bottom	Bread
RAT	UP	BILL		FOND	ROOT
Butter	Candle	Car		Charity	Cheese
FLY	BELL	BOY		FAITH	MITE
Claret	Curry	Egg		Ferry	For
RECTAL	COMB	PLANT		MAN	POUR
Game	Godfather	Harry		Honey	Jet
WILD	PATE	TOM		TRAP	LAG
Lake	Large	Lemon		Letter	Light
SEE	GRAND	MELON		BRIEF	HELL
Match	Measure for Measure	Merry Wives of Windsor		Midsummer Night's Dream	Monkey
GAME	ELBOW	SLENDER		COBWEB	SINGE
Much Ado About Nothing	Ochre	Oven		Pea	Peach
HERO	CHORE	FOUR		IRON	CHEAP
Pericles	Plane	Poison		Pond	Red
LEONINE	TREE	GIFT		MARE	ROT
Rose	Russet	Sausage		Ship	Silver
SORE	SUREST	ROLL		LAP	LIVERS
Sinker	Taming of the Shrew	Tan		Tempest	Throat
HOOK	PEDANT	ANT		IRIS	EAR
Ticket	Tomato	Tower	**Truck**	Twelfth Night	Van
BILLET	BACON	TOUR	**SYSTEM**	BELCH	GUARD

Themes are:

German/English: RAT/*ADVICE*, WILD/*GAME*, PATE/ *GODFATHER*, SEE/*LAKE*, BRIEF/*LETTER*, HELL/*LIGHT*, GIFT/ *POISON*, ROT/*RED*

A(, B &) C: UP(, up &) *AWAY*, BELL(, book &) *CANDLE*, FAITH(, hope &) *CHARITY*, TOM(, Dick &) *HARRY*, GAME(, set &) *MATCH*, HOOK(, line &) *SINKER*, EAR(, nose &) *THROAT*, BACON(, lettuce &) *TOMATO*

Can be preceded by mode of transport: *BOAT*BILL, *CAR*BOY, *FERRY*MAN, *JET* LAG, *PLANE* TREE, *SHIP*LAP, **TRUCK SYSTEM**, *VAN*GUARD

French/English: FOND/*BOTTOM*, POUR/*FOR*, GRAND/*LARGE*, SINGE/*MONKEY*, FOUR/*OVEN*, MARE/*POND*, BILLET/ *TICKET*, TOUR/*TOWER*

Can be preceded by a foodstuff: *BREAD* ROOT, *BUTTER* FLY, *CHEESE* MITE, *CURRY* COMB, *EGG* PLANT, *HONEY* TRAP, *PEA* IRON, *SAUSAGE* ROLL

Anagram of a colour: RECTAL/*CLARET*, MELON/*LEMON*, CHORE/*OCHRE*, CHEAP/*PEACH*, SORE/*ROSE*, SUREST/ *RUSSET*, LIVERS/*SILVER*, ANT/*TAN*

Shakespeare character/play (less definite and indefinite article): ELBOW/*MEASURE FOR MEASURE*, SLENDER/*MERRY WIVES OF WINDSOR*, COBWEB/*MIDSUMMER NIGHT'S DREAM*, HERO/*MUCH ADO ABOUT NOTHING*, LEONINE/*PERICLES*, PEDANT/*TAMING OF THE SHREW*, IRIS/*TEMPEST*, BELCH/ *TWELFTH NIGHT*

225.

Dewbius scored **two** penalties.

Countaphyt must have lost both their losing games 0-8 (to Apocrifal and Beaugas), as they gained no bonus points for losing by 7 or less. Hence they beat Dewbius 9-0.

Beaugas scored a total of 20 points against Apocrifal and Dewbius, and got one bonus point. This must have come from one of:

• Losing by 7 or less to Apocrifal. Not possible as Apocrifal could only have scored a maximum of 27 against them, which means Dewbius must have scored at least 51−27=24, in which case Dewbius must have beaten Beaugas.

• Scoring 4 tries against Apocrifal. Not possible as would have meant scoring 0 against Dewbius.

• Scoring 4 tries against Dewbius. This must have happened, and Beaugas must have scored 20 against Dewbius. They therefore scored 0 against Apocrifal.

Apocrifal must have scored 64−20−9=35 against Dewbius (as Dewbius had 64 points against them), therefore they scored 83−8−35=40 against Beaugas. Similarly Dewbius must have scored 24−0−0=24 against Apocrifal. Hence they scored 35−24−0=11 against Beaugas.

So Dewbius got their bonus point against Apocrifal, so must have scored 4 tries (2 converted). Hence they scored no penalties in that game. They scored 11 points against Beaugas, which can only come from a try and 2 penalties.

Overall scores were: Apocrifal beat Beaugas 40-0; Apocrifal beat Countaphyt 8-0, Apocrifal beat Dewbius 35-24; Beaugas beat Countaphyt 8-0, Beaugas beat Dewbius 20-11; Countaphyt beat Dewbius 9-0.

226.

(a) Units of time – Second, Minute, Hour, Day, Week, Month, Year, Decade, Century, Millennium: **S**, M, **H**, D, **W**, M, **Y**, D, **C**, M

(b) Atomic numbers of chemical elements not sharing first letter with symbol – Sodium (Na), Potassium (K), Iron (Fe). Silver (Ag), Tin (Sn), Antimony (Sb), Tungsten (W), Gold (Au), Mercury (Hg), Lead (Pb): 11, 19, 26, 47, 50, 51, 74, 79, **80, 82**

227.

Every other letter of countries, and the corresponding international dialling codes:

CANADA = 1, NORWAY = 47, CHINA = 86, CYPRUS = 357, **RUSSIA = 7**

228.

Names derived from anagrams:

(a) **Dan Abnormal** (anagram of Damon Albarn)

(b) **Vivian Darkbloom** (anagram of Vladimir Nabokov)

(c) **Torchwood** (anagram of Doctor Who)

(d) **Enid Coleslaw** (anagram of Daniel Clowes)

(e) **Narcotic Thrust** (anagram of Stuart Crichton)

229.

MOON.

1: UNIT: MOON UNIT is the elder daughter of Frank Zappa.

2: COMING, LIFE

3: TIMES, A, LADY

4: Preceded by FORE: ARM, SEE, SHORE, WARN

5: The: ALEXANDER (BORODIN), CESAR (CUI), MILY (BALAKIREV), MODEST (MUSSORGSKY), NIKOLAI (RIMSKY-KORSAKOV)

6: Contains SIX backwards: ALEXIS, AXIS, EXISTENCE, MARXIST, SEXISM, TAXIS

7: Leagues: BIG league, COLleague, DIFFERENT league, HANSEATIC league, PREMIER league, RUGBY league, SUPER league

8: Can be followed by -ATE (and of lengths 2–9): ALIENate, AUTHENTICate, CANDIDate, INNate, INTERPOLate, NEONate, ORate, PASSIONate

9: Ladies Dancing: LABEL/**B**ELLA, GALENA/**A**NGELA, DAILY/**L**YDIA, AURAL/**L**AURA, DEAN/**E**DNA, SORE/**R**OSE, AID/**I**DA, OMANI/**N**AOMI, LEANER/**A**RLENE – BALLERINA is spelled by the initial letters of the ladies' names

10: Synonym begins TEN: CANOPY/TENT, INCLINE/TEND, LESSEE/TENANT, OCCUPATION/TENURE, PRINCIPLE/TENET, SENSITIVE/TENDER, SHOOT/TENDRIL, SINEW/TENDON, STRAINED/TENSE, STUBBORN/TENACIOUS

230.

ZERO.

JSYV = FOUR + 4 (each letter advanced by 4)

KNAJ = FIVE + 5

MQOPB = EIGHT + 8

POF = ONE + 1

VYQ = TWO + 2

WKUHH = THREE + 3

WRWN = NINE + 9

YOD = SIX + 6

ZERO = ZERO + 0

ZLCLU = SEVEN +7

231.

The odd word out is **HOME** (which is 'in'!). Pairings are:

BOXING/SILENT (precede DAY and NIGHT)

DOWN/UP

ERGO/OGRE (reversals of each other)

GROG/TILT (G+T=26, R+I=26, O+L=26; A=1, B=2, ..., Z=26)

HISS/TOMTOM (............ and - - - - - - - - - - - - in Morse code)

MAIL/WASH (can be prefixed with BLACK and WHITE)

OPPO/SITES

RATTLE/REFORM (anagrams of LATTER and FORMER)

232.

L is missing, appropriate for NOEL! The themes for the other letters of the alphabet are:

AEGHP (NATO phonetic alphabet) – (ALPHA)BET, (ECHO) LOCATION, (GOLF) BALL, (HOTEL) LOBBY, (PAPA) SMURF

BDTXZ (music artistes) – PLAN (B), TENACIOUS (D), ICE(-T), GENERATION (X), JAY(-Z)

JMNRS (music artistes) – (J) LO, (M) PEOPLE, ('N) SYNC, (R.) KELLY, (S) CLUB

FKOVW (chemical element symbols) – FLU(ORINE),
POT(ASSIUM), OX(YGEN), VAN(ADIUM), TUN(GSTEN)

CIQUY (homophones of synonyms) – (SEE) OBSERVE, (AYE)
YES, (CUE) HINT, (YEW) TREE, (WHY) RATIONALE

233.

Two solutions depending on whether the element/planet at 4 down is
chosen to be Mercury or Krypton (the home planet of Superman).
The sum of 1 down is 40 + 8 = **48**.

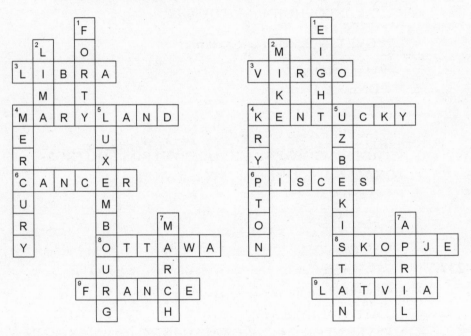

234.

(a) **WAX**. Scrabble scores increase. WAX scores 13.

(b) **POIKILOCYTE**. Increasing length, with vowels only in prime
positions. P-OI-K-I-L-O-CYT-E.

(c) **CLUE**. The sums of the letter values increases by 1. C+L+U+E=
3+12+21+5=41.

(d) **SLEEP**. Each word can be advanced by 1, 2, 3 . . . letters to form
a new word. SLEEP + 9 = BUNNY.

(e) **BEEF**. Each word can be advanced by 0, 1, 2 letters to form a
word when written backwards. BEEF + 10 = LOOP → POOL

235.

February 2nd. All the other dates fall on the same day of the week (Monday in 2016); it always falls on a different day (Tuesday in 2016).

236.

The phrase is SECRETARY GENERAL OF THE UNITED NATIONS and the list contains the post holders (or, more correctly, those elected to the post – hence the omission of GLADWYN JEBB) enciphered with the cipher phrase SECRTAYGNLOFHUIDBJKMPQVWXZ.

1. TRYGVE LIE (+1 letter in cipher phrase)

2. DAG HAMMARSKJOLD (+2)

3. **B GDNFG/U THANT (+3)**

4. KURT WALDHEIM (+4)

5. JAVIER PEREZ DE CUELLAR (+5)

6. **VBMOLBY VBMOLBY-UKFDP/BOUTROS BOUTROS-GHALI (+6)**

7. KOFI ANNAN (+7)

8. BAN KI-MOON (+8)

237.

Warshawski. Read the numbers as roman numerals.

EYE LASH (1=I), AYE AYE CAPTAIN (2=II), IVY LEAGUE (4=IV), VEE NECK (5=V), V. I. <u>WARSHAWSKI</u> (6=VI), EXPLAIN (10=X), EXCITED (11=XI).

238.

Which items (in the parenthesised sets) form the beginning of the name of a US state?

INDIA(NA), KENT(UCKY), RHO(DE ISLAND), TEN(NESSEE)

239.

(a) ISO country digraphs as Roman numerals:

$$\frac{MM+CM+ML+CC-CL}{LV+VI-LI} = \frac{2000+900+1050+200-150}{55+6-51} = 400 = CD = \textbf{Dem. Rep. of the Congo}$$

(b) ISO country digraphs as chemical symbols:

$$\frac{Mg\times((Au+Ni)-(Se+Cu))}{Na} = \frac{12\times((79+28)-(34+29))}{11} = 48 = Cd = \textbf{Dem. Rep. of the Congo}$$

(c) ISO country digraphs as hexadecimal numbers:

$$\frac{AF\times(BF+AF-CA)}{(DE-BB)\times(BE-BA)} = \frac{175\times(191+175-202)}{(222-187)\times(190-186)} = 205 = CD = \textbf{Dem. Rep. of the Congo}$$

(d) Treating letters as A=0, B=1, C=2, etc., do addition modulo 26 on each letter:

    i. Br + Kr = $(1,17) + (10,17) = (11,34) = (11,8) = Li = $ **Lithium**

    ii. Zn + Mg − Tc = $(25,13) + (12,6) - (19,2) = (18,17) = Sr$ **=Strontium**

(e) The number of days from birth date to death date for each person:

$$\frac{(25982 - 25974) \times (12137 - 12134) \times 26299}{(25988 - 25971) \times (26315 - 26314)} = 37128 = \textbf{The Queen Mother}$$

240.

First letters of:

(a) **Super Bowl winners**: Green Bay Packers, Green Bay Packers, New York Jets, Kansas City Chiefs, Baltimore Colts, Dallas Cowboys, Miami Dolphins, Miami Dolphins, Pittsburgh Steelers, Pittsburgh Steelers, Oakland Raiders, Dallas Cowboys, Pittsburgh Steelers, Pittsburgh Steelers, Oakland Raiders, San Francisco

49ers, Washington Redskins, Los Angeles Raiders, San Francisco 49ers, Chicago Bears, New York Giants, Washington Redskins, San Francisco 49ers, San Francisco 49ers, New York Giants, Washington Redskins, Dallas Cowboys, Dallas Cowboys, San Francisco 49ers, Dallas Cowboys, Green Bay Packers, Denver Broncos, Denver Broncos, St Louis Rams, Baltimore Ravens, New England Patriots, Tampa Bay Buccaneers, New England Patriots, New England Patriots, Pittsburgh Steelers, Indianapolis Colts, New York Giants, Pittsburgh Steelers, New Orleans Saints, Green Bay Packers, New York Giants, Baltimore Ravens, Seattle Seahawks, New England Patriots, Denver Broncos;

(b) **US presidential (birth) names**: George Washington, John Adams, Thomas Jefferson, James Madison, James Monroe, John Quincy Adams, Andrew Jackson, Martin Van Buren, William Henry Harrison, John Tyler, James Knox Polk, Zachary Taylor, Millard Fillmore, Franklin Pierce, James Buchanan, Abraham Lincoln, Andrew Johnson, Hiram Ulysses Grant, Rutherford Birchard Hayes, James Abram Garfield, Chester Alan Arthur, Stephen Grover Cleveland, Benjamin Harrison, Stephen Grover Cleveland, William McKinley, Theodore Roosevelt, William Howard Taft, Thomas Woodrow Wilson, Warren Gamaliel Harding, John Calvin Coolidge, Herbert Clark Hoover, Franklin Delano Roosevelt, Harry S. Truman, Dwight David Eisenhower, John Fitzgerald Kennedy, Lyndon Baines Johnson, Richard Milhous Nixon, Leslie Lynch King, James Earl Carter, Ronald Wilson Reagan, George Herbert Walker Bush, William Jefferson Clinton, George Walker Bush, Barack Hussein Obama

(c) **Monopoly squares**: Go, Old Kent Road, Community Chest, Whitechapel Road, Income Tax, King's Cross Station, The Angel Islington, Chance, Euston Road, Pentonville Road, In Jail/Just Visiting, Pall Mall, Electric Company, Whitehall, Northumberland Avenue, Marylebone Station, Bow Street, Community Chest, Marlborough Street, Vine Street, Free Parking, Strand, Chance, Fleet Street, Trafalgar Square, Fenchurch Street Station, Leicester Square, Coventry Street, Water Works, Piccadilly, Go To Jail, Regent Street, Oxford Street, Community Chest, Bond Street, Liverpool Street Station, Chance, Park Lane, Super Tax, Mayfair

(d) **Shipping forecast areas**: Viking, North Utsire, South Utsire, Forties, Cromarty, Forth, Tyne, Dogger, Fisher, German Bight, Humber, Thames, Dover, Wight, Portland, Plymouth, Biscay, Trafalgar, FitzRoy, Sole, Lundy, Fastnet, Irish Sea, Shannon, Rockall, Malin, Hebrides, Bailey, Fair Isle, Faeroes, Southeast Iceland

(e) **(Official) James Bond films**: Dr. No, From Russia with Love, Goldfinger, Thunderball, You Only Live Twice, On Her Majesty's Secret Service, Diamonds Are Forever, Live and Let Die, The Man with the Golden Gun, The Spy Who Loved Me, Moonraker, For Your Eyes Only, Octopussy, A View to a Kill, The Living Daylights, Licence to Kill, GoldenEye, Tomorrow Never Dies, The World Is Not Enough, Die Another Day, Casino Royale, Quantum of Solace, Skyfall, Spectre

241.

(a) **NAIVELY**: the 14 7-letter words can be formed from the 98 non-blank tiles in an English Scrabble set.

(b) **RADON** (86): it has no letters in common with EIGHTY SIX. Similarly for HELIUM (2), BORON (5), CARBON (6), SODIUM (11), PHOSPHORUS (15), CALCIUM (20), IRIDIUM (77), GOLD (79).

(c) **99** (NINETY NINE): all other numbers contain, when written in English, two letters which are consecutive in the alphabet (but not necessarily consecutive in the word(s)).

242.

    (a) **Toni Basil**

    (b) **Anne Rice**

    (c) **Ben Cross**

    (d) **Leon Uris**

    (e) **Tim Roth**

    (f) **Carl Andre**

    (g) **Sean Bean**

    (h) **Jet Li**

    (i) **Midge Ure**

    (j) **Lou Reed**

243.

    (a) **14** (XIV) and **4050** (MMMML). The numbers have, when expressed in Roman numerals, Scrabble scores 1, 2, 3, 4, 5, 6, …

    (b) **13** (THIRTEEN) and **10^{19}** (or **10^{13}**) (TEN TRILLION). The numbers have, when expressed as cardinals in English, Scrabble scores 1, 2, 3, 4, 5, 6, … ('-' denotes an unachievable score).

    (c) **3** (Li) and **117** (Ts). The numbers, when converted to the symbol of the corresponding element, Scrabble scores 1, 2, 3, 4, 5, 6, … ('-' denotes an unachievable score).

244.

5, 4, 5, 4. A team's score is the number of letters its name has contained in the group title; repeated letters in a team's name must be repeated in the title to further count (which is why GREECE scores 3 for the Es and CZECH REPUBLIC scores only 1 for the Cs).

SCANDINAVIA				SOUTHERN EUROPE			
NORWAY	2	DENMARK	3	ITALY	1	SPAIN	3
F*INL*A*ND*	5	SWEDEN	3	FRANCE	3	G*REE*C*E*	4

CENTRAL EUROPE				REST OF THE WORLD			
*POLAN*D	5	CZECH REPUBLIC	7	WALES	4	I*SRAE*L	4
ROMANIA	4	RUSSIA	3	SOUTH AFRICA	6	ENGLAND	3

245.

Marriages:

(a) Gary Oldman – **Uma Thurman** – Ethan Hawke

(b) Michael Jackson – **Lisa Marie Presley** – **Nicolas Cage** – Patricia Arquette

(c) David Frost – **Lynne Frederick** – **Peter Sellers** – **Britt Ekland** – Slim Jim Phantom

(d) Debbie Reynolds – **Eddie Fisher** – **Elizabeth Taylor** – **Michael Wilding** – **Margaret Leighton** – Laurence Harvey

(e) Lex Barker – **Lana Turner** – **Artie Shaw** – **Ava Gardner** – **Frank Sinatra** – **Mia Farrow** – André Previn

246.

(a) **ABBA** = Agnetha Faltskog, Anni-Frid Lyngstad, Benny Andersson, Björn Ulvaeus

(b) **Beatles** = George Harrison, John Lennon, Paul McCartney, Ringo Starr

(c) **Cream** = Eric Clapton, Ginger Baker, Jack Bruce

(d) **Doors** = John Densmore, Jim Morrison, Robby Krieger, Ray Manzarek

(e) **Eurhythmics** = Annie Lennox, Dave Stewart

(f) **Fugees** = Lauryn Hill, Pras Michel, Wyclef Jean

(g) **Genesis** = Mike Rutherford, Phil Collins, Peter Gabriel, Steve Hackett, Tony Banks

(h) **Hanson** = Clarke Hanson, Jordan Hanson, Zachary Hanson

(i) **INXS** = Andrew Farriss, Garry Beers, Jon Farriss, Kirk Pengilly, Michael Hutchence, Tim Farriss

(j) **Jam** = Bruce Foxton, Paul Weller, Rich Buckler

(k) **Kiss** = Ace Frehley, Gene Simmons, Peter Criss, Paul Stanley

(l) **Lindisfarne** = Alan Hull, Rod Clements, Ray Jackson, Ray Laidlaw, Simon Cowe

(m) **Madness** = Chris Foreman, Carl Smyth, Daniel Woodgate, Graham McPherson, Lee Thompson, Mike Barson, Mark Bedford

(n) **Nirvana** = Dave Grohl, Kurt Cobain, Krist Novoselic

(o) **Oasis** = Liam Gallagher, Noel Gallagher, Paul Arthurs, Paul McGuigan, Tony McCarroll

(p) **Pulp** = Candida Doyle, Jarvis Cocker, Nick Banks, Russell Senior, Steve Mackey

(q) **Queen** = Brian May, Freddie Mercury, John Deacon, Roger Taylor

(r) **REM** = Bill Berry, Mike Mills, Michael Stipe, Peter Buck

(s) **Squeeze** = Chris Difford, Gilson Lavis, Glenn Tilbrook, Harry Kakoulli, John Bentley

(t) **Temptations** = David Ruffin, Eddie Kendricks, Melvin Franklin, Paul Williams, Otis Williams

(u) **U2** = Adam Clayton, David Evans, Larry Mullen, Paul Hewson

(v) **Verve** = Nick McCabe, Peter Salisbury, Richard Ashcroft, Simon Jones

(w) **Who** = John Entwistle, Keith Moon, Peter Townshend, Roger Daltrey

(x) **XTC** = Andy Partridge, Barry Andrews, Colin Moulding, Terry Chambers

(y) **Yazoo** = Alison Moyet, Vince Clarke

(z) **Zutons** = Abi Harding, Boyan Chowdhury, David McCabe, Russell Pritchard, Sean Payne

247.

DEN (Ounce) fits between GLASS (Onion) and BOFFIN (Our Mutual Friend).

Aciform	Afghanistan	Austin	Back	Barnaby Rudge
NEEDLE	BAULK	MAESTRO	GET	DAISY
Bleak House	Bolivia	Chain	Cruciform	Cuneiform
BADGER	CURSE	FOOD	CROSS	WEDGE
David Copperfield	Dog	Dombey and Son	Ensiform	Falciform
STRONG	SALTY	CHICK	SWORD	SICKLE

Ford	France	Gin	Inch	Italy
TELSTAR	PAIRS	PINK	CHAFF	MORE
Julep	Lagoon	Lane	Little Dorrit	Lotus
MINT	BLUE	PENNY	TICKET	ELITE
Man	Martin Chuzzlewit	Mary	Mule	Napiform
NOWHERE	PINCH	BLOODY	MOSCOW	TURNIP
Nicholas Nickleby	Nissan	Norway	Onion	Ounce	Our Mutual Friend
HAWK	CHERRY	SOLO	GLASS	**DEN**	BOFFIN
Peru	Pint	Pisiform	Pole	Pound
MAIL	CUCKOO	PEA	HOP	DOG
Prudence	Renault	Retiform	Robert	Russian
DEAR	MEDALLION	NET	DOCTOR	WHITE
South Korea	Stone	Together	Toyota	Tripper
LOUSE	BRIM	COME	DUET	DAY
Tunisia	Vauxhall	Velvet	Volkswagen	Yard
UNITS	INSIGNIA	BLACK	BEETLE	BARN

Themes are:

Object/shape adjective (or English/Latin): NEEDLE/*ACIFORM* (Acus), CROSS/*CRUCIFORM* (Crux, Crucis), WEDGE/ *CUNEIFORM* (Cuneus), SWORD/*ENSIFORM* (Ensis), SICKLE/ *FALCIFORM* (Falx, Falcis), TURNIP/*NAPIFORM* (Napus), PEA/ *PISIFORM* (Pisum), NET/*RETIFORM* (Rete)

Anagram of capital of a country: BAULK/*AFGHANISTAN* (Kabul), CURSE/*BOLIVIA* (Sucre), PAIRS/*FRANCE* (Paris), MORE/*ITALY* (Rome), SOLO/*NORWAY* (Oslo), MAIL/*PERU* (Lima), LOUSE/ *SOUTH KOREA* (Seoul), UNITS/*TUNISIA* (Tunis)

Car manufacturer/model: *AUSTIN* MAESTRO, *FORD* TELSTAR, *LOTUS* ELITE, *NISSAN* CHERRY, *RENAULT* MEDALLION, *TOYOTA* DUET, *VAUXHALL* INSIGNIA, *VOLKSWAGEN* BEETLE

Beatles' songs – first/last word: GET *BACK*, PENNY *LANE*, NOWHERE *MAN*, GLASS *ONION*, DEAR *PRUDENCE*, DOCTOR *ROBERT*, COME *TOGETHER*, DAY *TRIPPER*

Surname/Dickens' book where it occurs: DAISY/*BARNABY RUDGE*, BADGER/*BLEAK HOUSE*, STRONG/*DAVID COPPERFIELD*, CHICK/*DOMBEY AND SON*, TICKET/*LITTLE DORRIT*, PINCH/ *MARTIN CHUZZLEWIT*, HAWK/*NICHOLAS NICKLEBY*, BOFFIN/*OUR MUTUAL FRIEND*

Can be followed by an Imperial unit: FOOD *CHAIN*, CHAFF*INCH*, **DENOUNCE**, CUCKOO*PINT*, HOP *POLE*, DOG *POUND*, BRIM*STONE*, BARN*YARD*

Cocktails – first/last word: SALTY/*DOG*, PINK/*GIN*, MINT/*JULEP*, BLUE/*LAGOON*, BLOODY/*MARY*, MOSCOW/*MULE*, WHITE/ *RUSSIAN*, BLACK/*VELVET*

248.

ITCHY.

1 ICHI (いち) – Japanese for 'one' – homophone: ITCHY

2 DEW + WET pronounced DUET: DEW, WET

3 Can be preceded by TRI-: AGE, BE, VIA

4 Associated with ivy (IV=4): COMMON, GROUND, LEAGUE, POISON

Can be followed by 5: BABYLON, CHANNEL, FAMOUS, SATURN, SLAUGHTERHOUSE

6 Synonyms of CEASE (French SIX/6): CONCLUDE, END, FINISH, HALT, STOP, TERMINATE

7 oaks: BEAR, BLUFF, PIN, POST, TURKEY, VALLEY, WATER

Pieces of 8: EGG, EIGHTIETH, GHEE, GIG, TEETH, THEE, TIGHT, TITHE

9 lives: AFTER, GOOD, LONG, LOVE, NIGHT, REAL, SHELF, STILL, WILD

Scrabble score of 10: BLOWN, EGOTISM, GHETTO, JUT, KITTEN, LORDLY, MORNING, SIX, THISTLE, TIPSY

249.

Extract the letters A–G from the opening lines of these songs and use them as notes to form the tune, retaining the original rhythm of the song.

(a) 'Away in a Manger' – AwAy in A mAnGEr, no CriB For A BED

(b) 'Deck the Halls' – DECk thE hAlls with BouGhs oF holly, FA lA lA lA lA lA lA lA lA

(c) 'The Twelve Days of Christmas' – on thE First DAy oF ChristmAs my truE lovE GAvE to mE, A pArtriDGE in A pEAr trEE

(d) 'Gaudete' – GAuDEtE, GAuDEtE, Christus Est nAtus, Ex mAriA virGinE, GAuDEtE

250.

(a)				(b)			
AFGHANISTAN	2	GABON	4	SOFIA	3	BISHKEK	2
PORTUGAL	5	AZERBAIJAN	4	SUVA	3	SUCRE	5

In (a) FOUR is spelled out by the nth (n = score) letter of the country's name; in (b) FIVE is spelled out by the nth letter of the country's capital.

251.

(a) BERYLLIUM/BE is the only element whose symbol's Scrabble score is equal to its atomic number.

(b) ALUMINIUM is the only element whose Scrabble score is equal to its atomic number.

(c) IRON/FE and **LEAD/PB** are the only elements where the

Scrabble score of the symbol exceeds that of the name.

(d) **TIN** has all its letters contained in most (16) other elements (AC*TIN*IUM, ANTIMONY, ASTA*TIN*E, EINSTEINIUM, MEITNERIUM, NEPTUNIUM, NITROGEN, PLA*TIN*UM, PLUTONIUM, PROTAC*TIN*IUM, ROENTGENIUM, RUTHENIUM, STRONTIUM, TECHNETIUM, TITANIUM, TENNESSINE) and is a sub-string of most (4 – see above) other elements. Either answer is acceptable.

(e) **LEAD** (is the only element that) contains no element symbol as a sub-string; **PROTACTINIUM** contains the most (AC, C, I, IN, N, NI, O, P, PR, TA, TI, U).

252.

The clue leads to:

Weaponry thought to damage, left on boat, dictates that I conquered German river.

ARMS   IDEA   HARM PORT   SAYS   VICI   RUHR

which can be written as:

AR<u>MS</u>

ID<u>EA</u>

HA<u>RM</u>

PO<u>RT</u>

SA<u>YS</u>

VI<u>CI</u>

RU<u>HR</u>

The last two columns spell out **Merry Christmas**.

253.

(a) **163.** The sequence is the square numbers written in reverse, starting with 144. 361 reversed is 163.

(b) **50.** The sequence is those numbers which are the sum of two squares. 50 is the first number that is the sum of two squares in two different ways. 50 = 49 + 1 and 50 = 25 + 25.

254.

(a) Twelve Days of Christmas:

$$\frac{\text{Drummers Drumming} + \text{Ladies Dancing} + \text{Turtle Doves}}{\text{Lords a Leaping} + \text{Swans a Swimming} + \text{Geese a Laying}} = \frac{12 + 9 + 2}{10 + 7 + 6}$$

= 1 = Partridge in a Pear Tree = **PiaPT**

(b) Monetary values in Monopoly:

$$\frac{(£100 \times £100) + (£50 \times -£230)}{£10} = -£150 = \textbf{School fees}$$

An alternative answer would be to use the general hotel repair penalty of -£100 rather than the street repair figure of -£115. This would lead to an overall answer of £0, but there isn't an obvious way to represent this as a value in Monopoly.

(c)

$$\frac{\text{Five Points}}{\text{Three Rivers}} \times \frac{\text{Four Oaks}}{\text{Five Forks}} \times \frac{\text{Three Forks}}{\text{Three Points}} \times \frac{\text{Two Rivers}}{\text{Three Oaks}} \times \text{Ninety Mile Beach} =$$

Eighty Mile Beach = (19°57'S, 119°54'E) or thereabouts (it is 80 miles long!)

255.

(a) **D** – the sequence is the Roman numeral letters contained in the names of the months in French: JAN**VI**ER, FÉ**V**R**I**ER, **M**ARS, A**V**R**IL**, **M**A**I**, JU**I**N, JU**ILL**ET, AOÛT, SEPTE**M**BRE, O**C**TOBRE, NO**VEM**BRE, **DÉCEM**BRE

(b) **C** – the sequence is the Roman numeral letters contained in the names of the months of the French Revolutionary Calendar: **V**EN**DÉMI**AIRE, BRU**M**AIRE, FR**IM**AIRE, N**IVÔ**SE, PLU**VIÔ**SE, **V**ENTÔSE, GER**MIN**AL, FLORÉAL, PRA**IRI**AL, **M**ESS**ID**OR, THER**MID**OR, FRU**C**T**ID**OR

256.

Ordered by Chinese New Year creature:

Dog/2018, Dragon/2024, Goat/2027, Horse/2026, Monkey/2016, Ox/2021, Pig/2019, Rabbit/2023, Rat/2020, Rooster/2017, Snake/2025, Tiger/2022

257.

Precisely, the key is the description below. So, for example, for a:

```
PLAIN:   ABCDEFGHIJKLMNOPQRSTUVWXYZ
CIPHER:  TRIVALPUSCOBDEFGHJKMNQWXYZ
```

(a) **Trivial Pursuit colours**: ORANGE, PINK, GREEN, BLUE, BROWN, YELLOW

(b) **Cluedo weapons**: LEAD PIPE, DAGGER, REVOLVER, ROPE, SPANNER, CANDLESTICK

(c) **The Fellowship of the Ring**: PIPPIN, MERRY, BOROMIR, SAM, LEGOLAS, FRODO, ARAGORN, GIMLI, GANDALF

(d) **Dozen, baker's dozen, score, gross**: ONE HUNDRED AND FORTY FOUR, THIRTEEN, TWELVE, TWENTY

258.

(a) **Three choirs**: Gloucester, Hereford, Worcester

(b) **Four Gospels**: Matthew, Mark, Luke, John

(c) **Cinque Ports**: Hastings, New Romney, Hythe, Dover, Sandwich

(d) **Six members of Monty Python**: Chapman, Cleese, Gilliam, Idle, Jones, Palin

(e) **Seven Heavenly Virtues**: Faith, Hope, Charity, Fortitude, Justice, Temperance, Prudence

259.

(a) **APED, AT** (or any other word beginning with A that can be preceded by NE to form a word): H-AIR, HE-ART, LI-ABILITY, BE-AVER, B-ANGLE, C-AMBER, N-ACRE, O-AT, F-ACTOR, NE-APED

(b) **DIM** (or any other 3-letter word formed from the letters of SODIUM): DRY/HYDROGEN, LIE/HELIUM, HIT/LITHIUM, RYE/BERYLLIUM, ROB/BORON, BAR/CARBON, GET/NITROGEN, YEN/OXYGEN, NIL/FLUORINE, ONE/NEON, DIM/SODIUM

(c) **OIL** (or any other 3-letter word formed from the letters of KILO): LAP/ALPHA, ROB/BRAVO, AIL/CHARLIE, LAD/DELTA, HOE/ECHO, TOO/FOXTROT, FOG/GOLF, LET/HOTEL, DIN/INDIA, LIE/JULIET, OIL/KILO

(d) **KOP, KOS** (or any 3 letter word beginning with K and its letters in strict alphabetical order)

260.

NDH LY BXGYDNH/RIO DE JANEIRO (XXXI OLYMPIAD). The sequence is the list of cities that have hosted the Summer Olympics since 1960 (ROME, TOKYO, MEXICO CITY, MUNICH, MONTREAL, MOSCOW, LOS ANGELES, SEOUL, BARCELONA, ATLANTA, SYDNEY, ATHENS, BEIJING, LONDON) enciphered with the Olympiad designation (XVII OLYMPIAD, XVIII OLYMPIAD,. . ., XXX OLYMPIAD).

261.

(a) **THE FIFTH ELEMENT**, B being the symbol for Boron (atomic number 5)

(b) **TWELVE MONKEYS**, £6000 = 12×£500 and 'MONKEY' is slang for £500

262.

Map the QWERTYUIOP row on a keyboard to the digits in the row above in the standard way, using Q=1, W=2, ..., O=9, P=0. If a character doesn't appear in that row, use '.' instead. The sequence is the result of mapping ZERO, ONE, TWO etc:

```
ZERO, ONE, TWO, THREE, FOUR, FIVE, SIX
.349, 9.3, 529, 5.433, .974, .8.3, .8.
```

263.

(a) **Book/Film adapted from book, Author**:

    i.   The Body/Stand By Me, Stephen King

    ii. Schindler's Ark/Schindler's List, Thomas Keneally

    iii. Northern Lights/The Golden Compass, Philip Pullman

    iv. Red Dragon/Manhunter, Thomas Harris

    v.   Call for the Dead/The Deadly Affair, John le Carré

    vi. Do Androids Dream of Electric Sheep?/Blade Runner, Philip K. Dick

(b) Reduplicative expressions:

AB = Argy Bargy = dispute, HP = Hocus Pocus = trickery, NP = Namby Pamby = insipid

    i.   FD = Fuddy Duddy = **dull person**

    ii. WN = Willy Nilly = **without care**

    iii. **RP** = Roly Poly = pudding

    iv. **HG** = Hurdy Gurdy = musical instrument

(c) Names containing chemical elements:

barba**RA DON**nelly

ba**SIL VER**non

da**LE AD**ams

dou**G OLD**field

ja**NE O'**Neill

NICK ELlis

os**CAR BON**d

pat**TI N**orman

r**OXY GEN**try

tam**ZIN C**ooper

264.

DOZEN (contains ONZE):

AMUSING (UN), EXODUS (DEUX), HISTORY (TROIS), EQUATOR (QUATRE), QUICKEN (CINQ), SPHINX (SIX), OPPOSITE (SEPT), AZIMUTH (HUIT), EFFLUENT (NEUF), INDEX (DIX)

265.

They're the 10 points on the Mohs scale of mineral hardness:

TALC+1=UBMD, GYPSUM+2=IARUWO, CALCITE+3=FDOFLWH, FLUORITE+4=JPYSVMXI, APATITE+5=FUFYNYJ, ORTHOCLASE FELDSPAR+6=UXZNUIRGYK LKRJYVGX, QUARTZ+7=XBHYAG, TOPAZ+8=BWXIH, CORUNDUM+9=LXADWMDV, DIAMOND+10=NSKWYXN

266.

Entries in the crossword grid are from the periodic table of elements (mostly the chemical symbol, sometimes the element name, and sometimes both).

1 Silver	■	**2** Ar	**3** Se	Ni	**4** C	■	**5** I
Ba	■	**6** Ca	Po	■	**7** He	W	N
8 C	Ar	Ne	Y		Er	■	Fe
K	■	■		**9** He	Pb Lead	U	Rn
■	**10** He	**11** Li	Co	Pt	Er	■	O
12 C	■	Na	■	Ag	■	**13** P	■
14 Si	**15** S	■	**16** O	**17** As	I		**18** S
■	**19** P	Ar	**20** La	N	Ce	■	Ta
21 Fe Iron	Y	■	Os	■	Nd	■	Tin

267.

(a) **10000.** The sequence represents 16 in bases 10 down to 2.

(b) **121.** Squares in base 9.

(c) **A9.** Squares in hex.

(d) **10.** Each number is the previous one in hex.

(e) **14641.** N^4 base $(N-1)$. All the numbers in the rest of the sequence are 14641.

268.

(a) Old terms for golf irons. There are two answers depending on the interpretation of Mashie-Niblick in the question – is that a hyphen or a minus sign?

$\sqrt{(2^2 + 5^2 + 5\text{-}9\,)}$ = 5 = **Mashie**

$\sqrt{(2^2 + 5^2 + 7\,)}$ = 6 = **Spade Mashie**

(b) Morse numbers. There are two ways of parsing the first summand:

(-----).(---..) + ... = 0.8 + 5 = 5.8 = ...---..

-(----.---..)   + ... = -98 + 5 = -93 = -----....--

269.

(a) **NIB** [_N_-DUBZ; _I_ CLAUDIUS, ROBOT; DRONE, HONEY _B_EE, STING]

(b) **ZIP** [_Z_ULU; _I_-PAD, POD; GARDEN, SPLIT, SWEET _P_EA]

(c) **LOLLY** [E_L_-AL; EVIAN, PERRIER – EAU/_O_; E_L_APSE, E_L_ATE, E_L_BOW; E_L_EVEN, E_L_FIN, E_L_GAR, E_L_IDE; CHEPSTOW, HAY, HEREFORD, MONMOUTH, ROSS – all on the WYE/_Y_]

(d) **POINT** [CHICK/_P_EA; _O_ DEAR, NO; BLACK, EVIL, PRIVATE EYE/_I_; BRIGHT-, LENGTH-, SHORT-, STIFF-E_N_; _T_/TEA BAG, COSY, CUP, LEAF, POT]

(e) **JUICE** [_J_AY WALKING; _U_-BOAT, TURN; OBSERVE, OGLE, VIEW – synonyms of EYE/_I_; _C_/SEA – HORSE, LION, SHORE, WEED; _e_-BAY, COLI, CRIME, GOVERNMENT, MAIL]

(f) **NEWS** [E*N*-JOY; ATTEND-, LESS-*EE*; *W* – DUB BELL EWE;
AUTHOR-, COUNT-, GIANT-, PRIEST-ES*S*]

(g) **MEXICO** [BONEY *M*; *E* = EMCEE SQUARED; *X* MARKS
THE SPOT; *I* = SQUAREROOT OF MINUS ONE; LOTS
MORE FISH IN THE *C*/SEA; *O* FOR THE WINGS OF A
DOVE]

(h) **RETSINA** [*R*HO – DEATH; *E*PSILON – LESION, PONIES
(letters can be found in EPSILON); *T*AU – EVACUATE,
GUATEMALA, SQUAT (contain TAU reversed); *S*IGMA –
AGEISM, IMAGES, MAGICS, STIGMA (contain letters of
SIGMA); *I*OTA – BIT, SCINTILLA, SMIDGEON, SPECK,
WHIT; *N*U(NEW) – AGE, ENGLAND, GUINEA, LABOUR,
YEAR, YORK; *A*LPHA – BET, DECAY, MALE, NUMERIC,
PARTICLE, RADIATION, RHYTHM]

270.

It is oetfn ciameld taht a snteecne or prshae is utdsaednlnrbae eevn if olny the
It is often claimed that a sentence or phrase is understandable even if only the

fsrit and lsat ltetres of ecah wrod are in the rghit pcale and the rset jbumeld.
first and last letters of each word are in the right place and the rest jumbled.

u   ah   ubo   eh   h   rsi   n   sa   retet   r   sinsi   n   h   se   lit
But what about when the first and last letters are missing and the rest still

mbuel ?   ah   re   h   wollnio   ingillar ?
jumbled? What were the following originally?

(a) **One two three four five**

(b) **Supercalifragilisticexpialidocious**

(c) **God Save Our Gracious Queen**

(d) **Into each life some rain must fall**

(e) **All for one and one for all**

(f) **To be or not to be**

271.

TROMBONES.

	OPEN	BRACKETS	(
FOUR	FEATHERS		4
TIMES	TABLES		×
	LEGS	ELEVEN	11
	SIGN OF THE	TIMES	×
THREE	CARD TRICK		3
PLUS	FOURS		+
TWENTY	QUESTIONS		20
	CLOSE	BRACKETS	)
DIVIDED BY	A COMMON LANGUAGE		/
	IT TAKES	TWO	2
	FIRST AMONG	EQUALS	=
SEVENTY SIX	TROMBONES		76

272.

- (a) **L. Frank Baum**
- (b) **M. Night Shyamalan**
- (c) **J. Arthur Rank**
- (d) **L. Ron Hubbard**
- (e) **G. Gordon Liddy**
- (f) **F. Murray Abraham**
- (g) **F. Scott Fitzgerald**
- (h) **R. Daneel Olivaw** (from Isaac Asimov's robot stories)

273.

Pi. The description has been enciphered by adding the digits of pi letter by letter.

```
  THE RATIO OF THE CIRCUMFERENCE OF A CIRCLE TO ITS DIAMETER

+ 314 15926 53 589 7932384626433 83 2 795028 84 197 16939937

= WII SFCKU TI YPN JRUEXUJKTKRFH WI C JRWCNM BS JCZ EOJPNCHY
```

274.

ENTRY (No) fits between HIRE (New Hamps) and NEW (Nouveau).

Adams	Al	Alastor	Am	Argus
WOLFIE	GORE	MOODY	DRAM	FILCH
At	Au	Bartemius	Be	Beau
LAST	PAIR	CROUCH	PREPARED	BOYFRIEND
Bunker	Bureau	Caddie	Château	Color
ARCHIE	OFFICE	TEA	CASTLE	ADO
Connecti	Cornelius	Del	Do	Eagle
CUT	FUDGE	AWARE	DIE	GOLDEN
Fordham	Fore	Friend	Gâteau	Green
VIKING	PINA	FOE	CAKE	EVER
Hankey	He	Heads	In	Iron
COUNT	MAN	TAILS	LOVE	FLAT
Kist	Klaasen	Lavender	Make	Mary
LIPSTICK	MATADOR	BROWN	BREAK	LAND
More	New Hamps	**No** Nouveau	Plateau	Portmanteau
LESS	HIRE	**ENTRY** NEW	PLAIN	BAG
Putt	Remus	Sink	Sirius	Sooner
SHOT	LUPIN	SWIM	BLACK	LATER
Tableau	Taylor	Tennes	Thomas	Trick
SCENE	POWER	SEE	RIDDLE	TREAT
Van Barneveld	Washing	Webster	Wiscon	Wood
BARNEY	TON	SPIDER	SIN	GOOD

Themes are:

Darts players' nicknames: MARTIN *ADAMS*/WOLFIE, ANDY *FORDHAM*/THE VIKING, TED *HANKEY*/THE COUNT, CHRISTIAN *KIST*/THE LIPSTICK, JELLE *KLAASEN*/THE MATADOR, PHIL *TAYLOR*/THE POWER, RAYMOND *VAN BARNEVELD*/BARNEY, MARK *WEBSTER*/THE SPIDER

Preceded by 2-letter element symbol: *AL* GORE, *AM* DRAM, *AT* LAST, *AU* PAIR, *BE* PREPARED, *HE* MAN, *IN* LOVE, **<u>NO</u> ENTRY**

Harry Potter characters: *ALASTOR* MOODY, *ARGUS* FILCH, *BARTEMIUS* CROUCH, *CORNELIUS* FUDGE, *LAVENDER* BROWN, *REMUS* LUPIN, *SIRIUS* BLACK, *THOMAS* RIDDLE

French synonyms ending -EAU: BOYFRIEND/*BEAU*, OFFICE/*BUREAU*, CASTLE/*CHÂTEAU*, CAKE/*GÂTEAU*, NEW/*NOUVEAU*, PLAIN/*PLATEAU*, BAG/*PORTMANTEAU*, SCENE/*TABLEAU*

Precedes golfing term: ARCHIE *BUNKER*, TEA *CADDIE*, GOLDEN *EAGLE*, PINA*FORE*, EVER*GREEN*, FLAT *IRON*, SHOT *PUTT*, GOOD*WOOD*

Ending of US state: *COLORA*DO, *CONNECTIC*UT, *DELAWARE*, *MARYLAND*, *NEW HAMPS*HIRE, *TENNES*SEE, *WASHING*TON, *WISCON*SIN

X or Y: *DO* or DIE, *FRIEND* or FOE, *HEADS* or TAILS, *MAKE* or BREAK, *MORE* or LESS, *SINK* or SWIM, *SOONER* or LATER, *TRICK* or TREAT

275.

Five first names in the font corresponding to the surname:

JIM	[Courier]
DAWN	[French]
HENRI	[Matisse]
NORMAN	[Rockwell]
ANTONIO	[Vivaldi]

Five surnames in the font corresponding to the first name:

[Bradley]	WIGGINS
[Felix]	MENDELSSOHN
[Franklin]	D ROOSEVELT
[Georgia]	O'KEEFFE
[Kristen]	STEWART

276.

BART.

1 BAR-ONE-T (inserting ONE into BART turns the abbreviation into the full word)

2 Owls' call: T(W)O-WIT, T(W)O-WOO

3 Three-piece suit: JACKET, TROUSERS, WAISTCOAT

4 'FO(U)R' marriage vows: FOR BETTER, FOR WORSE; FOR RICHER, FOR POORER

5s: FIVES COURT, ETON FIVES, FIVES GLOVES, RUGBY FIVES, WINCHESTER FIVES

6×6 word square:

E	R	O	D	E	S
R	E	G	I	M	E
O	G	I	V	E	S
D	I	V	E	R	T
E	M	E	R	G	E
S	E	S	T	E	T

7 Swans(a-swimming): SWANAGE, SWAN DIVE, SWAN LAKE, SWANSEA, SWAN SONG, SWAN THEATRE, SWAN UPPING

Figure (of 8): CALCULATE, DIAGRAM, DIGIT, FORM, ILLUSTRATION, NUMBER, RECKON, SHAPE

9 Comic characters ending -IX: CACOPHONIX (BARD), VITALSTATISTIX (CHIEF), DOGMATIX (DOG), GETAFIX (DRUID), GERIATRIX (ELDER), UNHYGIENIX (FISHMONGER), ASTERIX (GAUL), HOMEOPATHIX (MERCHANT), FULLIAUTOMATIX (SMITH)

TEN- homophones: TEN-COMMAND-MINCE (TEN COMMANDMENTS), TEN-ERR-REEF (TENERIFE), TEN-GALLOW-GNAT (TEN-GALLON HAT), TEN-NESSIE-WILLIAMS (TENNESSEE WILLIAMS), TEN-NIECE-ELBOW (TENNIS ELBOW)

277.

(a) **SORROW**: FULLY

JOY: RIDE, STICK

GIRL: GUIDE, POWER, TALK

BOY: GEORGE, SCOUT, WONDER, ZONE

SILVER: ADO, LINING, PLATE, STONE, SURFER

GOLD: COAST, CREST, CUP, FINCH, FINGER, RUSH

SECRET: AGENT, GARDEN, POLICE, SANTA, SEVEN, SQUIRREL, SOCIETY

(b) **MAGPIE** (ONE FOR SORROW, TWO FOR JOY, ...)

278.

First words of books in a series (including prologues where present):

(a) The Lord of the Rings, by J. R. R. Tolkien

(b) The Harry Potter books, by J. K. Rowling

(c) The Chronicles of Narnia (in reading order), by C. S. Lewis

(d) The James Bond books, by Ian Fleming

(e) The Fifty Shades trilogy, by E. L. James

(f) The Alexandria Quartet, by Lawrence Durrell

(g) The Discworld books, by Terry Pratchett

(h)   The Raj Quartet, by Paul Scott

(i)   The Palliser novels, by Anthony Trollope

(j)   A Dance to the Music of Time, by Anthony Powell

279.

(a)   **CLIFF**. Reversing the order of the words, first and last letters spell CENTIGRADE and FAHRENHEIT.

(b)   **EARWIG**. First/last letters alternate to give BLACK/WHITE and RIGHT/WRONG.

(c)   **VAIN**. Words are anagrams of names starting A, B, C, etc., male/female alternating. VAIN is anagram of IVAN.

(d)   **UMPTEEN**. Words are anagrams of planets with one letter changed. UMPTEEN is an anagram of NEPTU_M_E.

(e)   **JOTTER**. Each word contains A, B, C, etc. followed by an animal. J-OTTER.

280.

(a)   T, T, S, S, RM, PM, **BM, T** (Tinker, Tailor, Soldier, Sailor, ...)

(b)   **B, B, B, B** (Earliest letter in names of months: JANUARY, FEBRUARY, ...)

(c)   **E, E, E** (Earliest letter in ONE , TWO, THREE, ..., EIGHT, NINE, TEN, ...)

(d)   **24** (Earliest letter in primes: TWO, THREE, FIVE, ...)

281.

(a)   **(101, 102, 103)**: the numbers can be written in the shape of a triangle, for example:

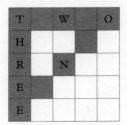

(b)   **(14, 15, 16, 17)**: the numbers can be written in the shape of a quadrilateral.

282.

H	Y	D	R	O	G	E
N	H	**E**	L	I	U	M
L	I	**T**	**H**	I	U	M
B	**E**	**R**	**Y**	**L**	L	I
U	M	B	**O**	**R**	**O**	N
C	A	R	B	O	N	N
I	**T**	R	O	G	**E**	N

(a) **BERYL** – thus referencing Conan Doyle's '(The Adventure Of) The Beryl Coronet'.

(b) It's a 7×7 table containing the first 7 elements.

283.

F.

THE KWIZ SETTERS ENJOY USING SIMPLE SUBSITITUTION CODES IN QUESTIONS. IN THIS EXAMPLE, HAD IT BEEN NECESSARY, WHICH LETTER WOULD HAVE BEEN USED FOR Z?

A	B	C	D	E	F	G	H	I	J	K	L	M	N	O	P	Q	R	S	T	U	V	W	X	Y	Z
T	H	E	K	W	I	Z	S	R	N	J	O	Y	U	G	M	P	L	B	C	D	Q	X	A	V	F

In the nth word of the text, the first n letters of the alphabet are enciphered: the cipher alphabet is derived from the order in which the letters first appear in the question.

284.

A	**Audi**	VDT	Vorsprung durch technik
B	**Bounty**	TTOP	The taste of paradise
C	**Castlemaine**	AWGAXFAE	Australians wouldn't give a XXXX for anything else
D	**Domestos**	KNNPOAKG	Kills ninety nine percent of all known germs

E	**Energizer**	IKGAGAG	It keeps going and going and going
F	**Ford**	EWDIDBY	Everything we do is driven by you
G	**Gilette**	TBAMCG	The best a man can get
H	**Heineken**	RTPOBCR	Refreshes the parts other beers cannot reach
I	**IBM**	SFASP	Solutions for a small planet
J	**Jaguar**	DDIDI	Don't dream it, drive it
K	**KitKat**	HABHAK	Have a break. Have a KitKat
L	**L'Oréal**	BIWI	Because I'm worth it
M	**Mars**	AMADHYWRAP	A Mars a day helps you work rest and play
N	**Nike**	JDI	Just do it
O	**Orange**	TFBTFO	The future's bright, the future's Orange
P	**Penguin**	PPPPUAP	P-P-P-Pick up a penguin
Q	**Qualcast**	IALLBTAH	It's a lot less bovver than a hover
R	**Rolo**	DYLAETGTYLR	Do you love anyone enough to give them your last Rolo?
S	**Schweppes**	SYKW	Schhh – you know who
T	**Typhoo**	YOGAOWT	You only get an oooo with Typhoo
U	**UPS**	MATSOB	Moving at the speed of business
V	**Volkswagen**	IOEILWARAAV	If only everything in life was as reliable as a Volkswagen
W	**Whiskas**	EOOTCPI	Eight out of ten cats prefer it
X	**Xerox**	TDC	The document company
Y	**Yell!**	LYFDTW	Let your fingers do the walking
Z	**Zanussi**	TAOS	The appliance of science

285.

$180 = (3 \times 5!)/(5-3)$ $181 = (35 \times 5) + 3!$

$182 = (5! \times .5r - 3!) \times 3$ $183 = 5!/.5r - 33$

$184 = 3!!/5 + 5! / 3$ $185 = 5 \times (5!/3 - 3)$

$186 = 5! + 33/.5$ $187 = 3/(.5^{3!}) - 5$

$188 = (5 - .3) \times 5!/3$ $189 = 5!/.5r - 3^3$

In fact, all the numbers from 1 to 192 can be constructed in this way. Doing this is an exercise for the reader!

286.

The first sum is correct as it is, but it remains correct when you square each of the terms in it:

$2+3+6+9 = 1+4+7+8$
$2^2+3^2+6^2+9^2 = 1^2+4^2+7^2+8^2$

The second sum has the same property, but also holds true if the terms are all cubed:

$2+3+10+11 = 1+5+8+12$
$2^2+3^2+10^2+11^2 = 1^2+5^2+8^2+12^2$
$2^3+3^3+10^3+11^3 = 1^3+5^3+8^3+12^3$

Finally

$2^n+3^n+9^n+13^n+16^n+20^n = 1^n+5^n+8^n+12^n+18^n+19^n$

holds true for n=1, 2, 3 and 4

The question then becomes: find such a sum that works for (at least) n=1, 2, 3, 4 and 5. For example:

1+6+7+17+18+23 = 2+3+11+13+21+22

287.

MORSE.

In the grid can be found the names of 20 animals, in French:

> ANE, ANGUILLE, CANARD, CERF, CHIEN, COCHON, CORBEAU, ESCARGOT, GRENOUILLE, GRILLON, GRUE, LION, OIE, ORIGNAL, PERROQUET, PINGOUIN, PUCE, RAT, SINGE, VER.

Remaining letters are:

```
A V E R C A N A R D
N A N I S W E C U P
G R E N O U I L L E
U E R S I S H F R R
I N O H C O C O C R
L E E N C A H R E O
L E G N I S R I R Q
E W O O A R E G F U
P I N G O U I N O E
L D C O R B E A U T
R A T G R I L L O N
```

These spell ANSWER IS FRENCH WORD, and they appear in the grid in a pattern that can be interpreted as Morse code (a single letter is a dot, a pair of letters is a dash). The pattern is:

-- --- .-. .... .

which in Morse code spells out MORSE. So, as indicated, the answer is the French word MORSE, which is French for Walrus.

288.

1:   Fairy (A fairy at the top of the tree)

2:   Proteus Valentine (Two Gentlemen of Verona)

3:   Aramis Athos Porthos (The Three Musketeers)

4:   Air Earth Fire Water (The four elements)

5:   Anne Dick George Julian Timmy (The Famous Five)

6:   A B D E E G (The standard tuning of a six-string guitar – E A D G B E)

7:   Bashful Doc Dopey Grumpy Happy Sleepy Sneezy (The Seven Dwarfs)

8:   Canada France Germany Italy Japan Russia United Kingdom United States (The G8 nations)

9:   Calliope Clio Erato Euterpe Melpomene Polyhymnia Terpsichore Thalia Urania (The Nine Muses)

10:   Eight Five Four Nine One Seven Six Ten Three Two (The numbers 1–10)

11:   Ball Banks Charlton Charlton Cohen Hunt Hurst Moore Peters Stiles Wilson (The eleven players in the 1966 World Cup winning England team)

12:   Apr Aug Dec Feb Jan Jul Jun Mar May Nov Oct Sep (The twelve months of the year)

13:   CT DE GA MA MD NC NH NJ NY PA RI SC VA (The thirteen original states of the Union)

Various alternative answers exist for the top of the tree. For example, for row 1,

    F = Franz (Franz von Werra – the One that got away)

    F = Formula (One)

Or for row 2, 'Peace, Victory' (2-fingered salutes).

289.

Two answers – atomic numbers:

$$\frac{(48 + 23 - 17)}{3} - \frac{48}{6} = 10 = \textbf{Neon}$$

Or chemical symbols interpreted as Roman numerals:

$$\frac{(CD + V - CL)}{LI} - \frac{CD}{C} = \frac{(400 + 5 - 150)}{51} - \frac{400}{100} = 1 = I = \textbf{Iodine}$$

290.

(a) **OVEREXCITABLE**. Attribute: increasing number of syllables.

(b) **OVEREXCITABLE**. Attribute: increasing number of vowels.

(c) **FLY**. Each word can be inserted into the next one. F-REE-LY.

(d) **DETAINS**. The number of letters in common between adjacent words increases by 1.

291.

AIAUAK. ELEVEN enciphered with the keyword ONZE by simple substitution: ABCDEFGHIJKLMNOPQRSTUVWXYZ → ONZEABCDFGHIJKLMPQRSTUVWXY; the sequence is enciphered likewise with keywords UN, DEUX, TROIS, ...

292.

K.

G	D	S	O	E	A	N	S	N
Q	N	T	K	H	M	B	E	U
R	I	I	I	A	R	E	D	A
E	E	L	N	M	O	F	F	M
H	A	R	L	I	S	O	O	N
T	F	S	Q	U	A	R	E	O
O	O	E	G	N	F	M	E	O
N	R	S	R	E	T	T	E	L
A	R	A	T	G	K	W	E	R

D	S	O	E	A	N	S
N	Q	T	K	H	M	B
E	U	I	I	A	R	E
D	A	E	L	M	O	F
A	R	L	S	N	F	O
O	E	G	N	E	O	R
R	A	T	G	K	W	E

D	O	A	N	N
T	H	E	I	D
E	M	A	S	N
O	G	N	E	O
A	T	G	K	W

A	N	D
O	N	E
T	G	K

| FORM ANOTHER SQUARE FROM REMAINING LETTERS | FORM SMALLER SQUARE LIKE BEFORE | NOW DO THE SAME AGAIN | AND ONE TO GO | K |

293.

(a) Cancer = **June** (Start of Ju-ly + end of Ju-ne)

(b) Routes with maximum number of stops on London Underground lines:

> Aldgate to Amersham = Metropolitan, Stanmore to Stratford = Jubilee, Cockfosters to Uxbridge = Piccadilly, High Barnet to Morden = Northern, Ealing Broadway to Uxminster = District, Epping to West Ruislip = Central, Brixton to Walthamstow Central = Victoria

So:

  i.   Barking to Hammersmith = Hammersmith and City = **H+C**

  ii. Bank to Waterloo = Waterloo and City = **W+C**

  iii. (Elephant+Castle) to (Harrow+Wealdstone) = Bakerloo = **B**

294.

All can be made into words by the addition of the letter 'E'.

BACON/BEACON, BOWLS/BOWELS, BURNT/BURNET, CANDID/CANDIED, CORPORAL/CORPOREAL, GALLON/ GALLEON, LAND/ELAND, POT/POET, QUALITY/ EQUALITY, QUIT/QUIET, RATIONAL/RATIONALE, RED/ REED, SILICON/SILICONE, SLIGHT/SLEIGHT, SPY/ESPY, SWINGING/SWINGEING, TRADING/TREADING, TWIN/ TWINE

295.

KWIZ. The remaining letters form a Scrabble set when the word squares below are formed.

a. WARP	e. JETS	i. CAFE	m. ROBE	q. THIS	u. GLUT
b. AQUA	f. EXAM	j. AVID	n. OILY	r. HIRE	v. LINO
c. RUIN	g. TACO	k. FIVE	o. BLUE	s. IRAN	w. UNDO
d. PANG	h. SMOG	l. EDEN	p. EYER	t. SEND	x. TOOT

296.

(a) **WILLIAM SHAKESPEARE**: The plays HAMLET, KING LEAR, THE TEMPEST, THE WINTER'S TALE, CORIOLANUS, TWELFTH NIGHT, TROILUS AND CRESSIDA, THE MERRY WIVES OF WINDSOR, THE MERCHANT OF VENICE, ANTONY AND CLEOPATRA with punctuation, spaces and letters in his name removed.

(b) **IAN LANCASTER FLEMING**: DR. NO, GOLDFINGER, MOONRAKER, CASINO ROYALE, LIVE AND LET DIE, THUNDERBALL, DIAMONDS ARE FOREVER, ON HER MAJESTY'S SECRET SERVICE, FROM RUSSIA, WITH LOVE, THE MAN WITH THE GOLDEN GUN with punctuation, spaces and letters in his name removed.

(c) **GEORGE ROBERT LAZENBY**: DR. NO (THOMAS SEAN CONNERY), CASINO ROYALE (DANIEL WROUGHTON CRAIG), LICENCE TO KILL (TIMOTHY PETER DALTON), GOLDENEYE (PIERCE BRENDAN BROSNAN), MOONRAKER (ROGER GEORGE MOORE), DIE ANOTHER DAY (PIERCE BRENDAN BROSNAN), QUANTUM OF SOLACE (DANIEL WROUGHTON CRAIG), TOMORROW NEVER DIES (PIERCE BRENDAN BROSNAN), THUNDERBALL (THOMAS SEAN CONNERY), FOR YOUR EYES ONLY (ROGER GEORGE MOORE), each film having punctuation, spaces and letters in the actor's name removed LAZENBY's ON HER MAJESTY'S SECRET SERVICE is unclued.

297.

507 (PANAMA) fits between 800 (OMEGA) and (PARIS) 297.

213 ALGERIA	659 ALUMINIUM	855 BENEDICT	320 BOND STREET		607 BONIFACE
673 BRUNEI	359 BULGARIA	850 CALCIUM	600 CHI		588 COSI FAN TUTTE
090 EAST	525 EINE KLEINE NACHTMUSIK	483 FELIX	731 GREGORY		366 IDOMENEO
361 JOHN	551 JUPITER	328 LEAD	795 LEO		423 LIECHTENSTEIN
425 LINZ	181 LITHIUM	650 MAGNESIUM	400 MAYFAIR		360 NORTH
045 NORTH-EAST	160 NORTHUMBERLAND AVENUE	315 NORTH-WEST	800 OMEGA	507 *PANAMA*	297 PARIS
120 PENTONVILLE ROAD	500 PHI	280 PICCADILLY	504 PRAGUE		700 PSI

626	100	200	962	432
REQUIEM	*RHO*	*SIGMA*	*SILVER*	*SIXTUS*
180	135	225	752	886
SOUTH	*SOUTH-EAST*	*SOUTH-WEST*	*STEPHEN*	*TAIWAN*
300	232	240	400	998
TAU	*TIN*	*TRAFALGAR SQUARE*	*UPSILON*	*UZBEKISTAN*
200	270	060	263	420
VINE STREET	*WEST*	*WHITECHAPEL*	*ZIMBABWE*	*ZINC*

Themes are:

International dialling codes: *ALGERIA*/213, *BRUNEI*/673, *BULGARIA*/359, *LIECHTENSTEIN*/423, **_PANAMA_**/**507**, *TAIWAN*/886, *UZBEKISTAN*/998, *ZIMBABWE*/263

Melting points of elements (Celsius): *ALUMINIUM*/659, *CALCIUM*/850, *LEAD*/328, *LITHIUM*/181, *MAGNESIUM*/650, *SILVER*/962, *TIN*/232, *ZINC*/420

Year Pope NAME III took up office: *BENEDICT*/855, *BONIFACE*/607, *FELIX*/483, *GREGORY*/731, *JOHN*/561, *LEO*/795, *SIXTUS*/432, *STEPHEN*/752

Monopoly board prices (£): *BOND STREET*/320, *MAYFAIR*/400, *NORTHUMBERLAND AVENUE*/160, *PENTONVILLE ROAD*/120, *PICCADILLY*/280, *TRAFALGAR SQUARE*/240, *VINE STREET*/200, *WHITECHAPEL*/060

Greek numbers: *CHI*/600, *OMEGA*/800, *PHI*/500, *PSI*/700, *RHO*/100, *SIGMA*/200, *TAU*/300, *UPSILON*/400

K-number of Mozart compositions: *COSI FAN TUTTE*/588, *EINE KLEINE NACHTMUSIK*/525, *IDOMENEO*/366, *JUPITER*/551, *LINZ (symphony)*/504, *PARIS (symphony)*/297, *PRAGUE (symphony)*/504, *REQUIEM*/626

Compass bearings: *EAST*/090, *NORTH*/360, *NORTH-WEST*/315, *NORTH-EAST*/045, *SOUTH*/180, *SOUTH-EAST*/135, *SOUTH-WEST*/225, *WEST*/270

298.

Each list has the letters in the description removed and is then ordered alphabetically.

(a) **SANTA'S REINDEER**: BLITZEN, DANCER, COMET, CUPID, DASHER, DONNER, PRANCER, VIXEN

(b) **(THE) FELLOWSHIP OF THE RING**: PIPPIN, LEGOLAS, ARAGORN, GANDALF, SAM, BOROMIR, FRODO, GIMLI, MERRY

299.

(a) **SALAAM**: starting at each of the 4 corners and moving horizontally or vertically (i.e. via Rook's moves) COD, STURGEON, MACKEREL and SALMON are spelled out.

(b) **BEAGLE**: similarly BACH, DEBUSSY, ELGAR and BEETHOVEN are spelled out.

(c) **CREEPS**: similarly PARAKEET, BLUETIT, SCOTER and CRANE are spelled out.

(d) **HORNET**: similarly MONTANA, DELAWARE, TENNESSEE and HAWAII are spelled out.

(e) **FRILLS**: similarly, but with Bishop's moves, CUBA, YEMEN, SENEGAL and FIJI are spelled out.

(f) **ABSORB**: similarly, but with King's moves, ARGENTINA, MACEDONIA, BARBADOS and AUSTRALIA are spelled out.

(g) **BOSBOK**: similarly, but with the appropriate style of moving, KNIGHT, ROOK, KING and BISHOP are spelled out.

300.

(a) **12451**. The list comprises the factors of 1 (1), 2 (1, 2), 3 (1, 3), 4 (1, 2, 4), ..., 19 (1, 19): the factors of 20 are (1, 2, 4, 5, 10, 20).

(b) **11010**. The list is the Roman numerals I, II, III, ..., (part of) XXI with I, V, X replaced by 1, 5, 10 respectively. XXI, XXI = 10101, 101011.

(c) **20521**. Cumulative totals traversing a dartboard clockwise from the '20': 20, 21 (=21+1), 39 (=20+1+18), ..., 193, 205, 210.

301.

101/1500.

3=LITHIUM=LI=51; 3/51=1/17

6=CARBON=C=100; 6/100=3/50

101=MENDELEVIUM=MD=1500; 101/1500

115=MOSCOVIUM=MC=1100; 115/1100=23/220

96=CURIUM=CM=900; 96/900=8/75

17=CHLORINE=CL=150; 17/150

48=CADMIUM=CD=400; 48/400=3/25

23=VANADIUM=V=5; 23/5

53=IODINE=I=1; 53/1=53

302.

Each given item from a category has at least one letter in common with all the other items in that category. For example, each (chemical) element contains one of the letters S, I, L, V, E, R.

303.

CHRISTMAS, the enciphered form of FIORDLAND. The other unenciphered penguins are: YELLOW-EYED, ROYAL, ROCKHOPPER, MAGELLANIC, MACARONI, AFRICAN, ADELIE, SNARES, KING, LITTLE, HUMBOLDT, GALAPAGOS, GENTOO, ERECT-CRESTED, EMPEROR, CHINSTRAP.

Answers to the second colour section

Colourful characters

  (a) **Justin Rose**

  (b) **P!nk**

  (c) **Erik the Red**

  (d) **Scarlett Johansson**

  (e) **Ginger Rogers**

  (f) **Jason Orange**

  (g) **Amber Heard**

  (h) **Gordon Brown**

  (i) **Ian Lavender**

  (j) **Violet Carson**

  (k) **Lionel Blue**

  (l) **Eva Green**

  (m) **Earl Grey**

  (n) **Jack Black**

  (o) **Betty White**

Next in sequence

Words ending in girls' names from the song 'Mambo No. 5':
har**MONICA**, am**ERICA**, na**RITA**. The next name is **TINA**, so the
answer could be a picture of Argentina, a Ford Cortina, a concertina,
etc.

Identify and divide into groups

  (a) **Rebecca Front**   (n) **Miranda Hart**

  (b) **Gladys Knight**   (o) **Chip**

  (c) **Johnny Cash**   (p) **Doris Day**

(d) **Dale**

(e) **Cary Grant**

(f) **Sean Kerly**

(g) **Wayne Rooney**

(h) **Blu Cantrell**

(i) **Chip**

(j) **Curly Howard**

(k) **Terry Waite**

(l) **Roger Black**

(m) **Michael Fish**

(q) **Mel C**

(r) **David Soul**

(s) **Clare Short**

(t) **Steve Waugh**

(u) **Sally Ride**

(v) **Ruby Wax**

(w) **Lauryn Hill**

(x) **Nick Park**

(y) **Heather Peace**

(z) **Neil Back**

The groups are as follows:

a and z: front and back

b and p: night and day

c and e: cash and carry

k and q: wait and see

l and h: black and blue

m and (i+o): fish and chips

n and r: heart and soul

s and (f+j): short and curlies

t and y: war and peace

v and g: wax and wane

w and d: hill and dale

x and u: park and ride

Identify the people

(a) **Charity Wakefield**

(b) **Chapman Pincher**

(c) **Sebastian Coe**

(d) **Jean Simmons**

(e) **Jeff Banks**

(f) **Penny Smith**

(g) **Harry Kewell**

(h) **Robin Tunney**

(i) **Phil Oakey**

(j) **Bryan Ferry**

(k) **Henrietta Maria of France** (wife of Charles I)

(l) **Ella Fitzgerald**

(m) **Cilla Black**

(n) **Georg Ohm**

(o) **Laurie Metcalf**

(p) **Allen Funt** (presenter of the US version of Candid Camera)

(q) **Geri Halliwell**

(r) **Les Bubb** (top British mime artist)

(s) **Carrie Fisher**

(t) **Jasper Carrott**

(u) **Steve Case** (former CEO of AOL)

(v) **Britney Spears**

(w) **Mr T**

(x) **Donald Driver**

(y) **Dee Dee Myers** (Press Secretary to President Clinton)

(z) **Molly Ringwald**

The pairings are such that if they married, the women would have unfortunate married names:

    a marries **u**: Charity Case

    d marries **n**: Jean Ohm

    f marries **b**: Penny Pincher

    h marries **e**: Robin Banks

    k marries **t**: Henrietta Carrott

    l marries **p**: Ella Funt

    m marries **r**: Cilla Bubb

    o marries **x**: Laurie Driver

    q marries **c**: Geri Coe

    s marries **i**: Carrie Oakey

    v marries **j**: Britney Ferry

    y marries **w**: Dee Dee T

    z marries **g**: Molly Kewell

Pictures

The items all feature in the 'Four Candles' sketch by the Two Ronnies. All the items are unwanted, except for the **13A Plug**.

Celebrity Sudoku

A9 = **Julia Roberts**   B2 = **John Major**   B7 = **Spencer Tracy**

C1 = **Kevin Sorbo**   C2 = **Pete Townshend**   C3 = **Mark Hamill**

D5 = **Sean Connery**   D6 = **Ernest Hemingway**   D8 = **Paul Weller**

D9 = **Marilyn Monroe**   E5 = **Rod Stewart**   E6 = **Marlon Brando**

E7 = **Elliot Gould**   F4 = **John Wayne**   F5 = **Dennis Taylor**

F7 = **Alan Sugar**   G1 = **Pierce Brosnan**   G2 = **David Schwimmer**

G4 = **Roy Hodgson**   G5 = **Katarina Witt**   G8 = **Steven Gerrard**

H2 = **Michelle Wie**   H6 = **Steven Spielberg**   H8 = **Kurt Cobain**

H9 = **Roger Bannister**   I1 = **Stella McCartney**   I4 = **Dizzy Gillespie**

The first initial of the name yields a Sudoku. The second initial yields another Sudoku.

								J
	J				S			
K	P	M						
			S	E		P	M	
			R	M	E			
			J	D	A			
P	D		R	K		S		
	M			S		K	R	
S			D					

								R
	M				T			
S	T	H						
			C	H		W	M	
			S	B	G			
			W	T		S		
B	S		H	W		G		
	W			S		C	B	
M		G						

Solving each Sudoku gives the following:

R	A	S	M	E	K	P	D	J
E	J	D	P	A	R	S	M	K
K	P	M	S	J	D	R	A	E
D	R	J	A	S	E	K	P	M
A	S	P	K	R	M	E	J	D
M	E	K	J	D	P	A	R	S
P	D	E	R	K	J	M	S	A
J	M	A	E	P	S	D	K	R
S	K	R	D	M	A	J	E	P

G	C	W	S	M	T	H	B	R
R	M	B	C	H	W	T	S	G
S	T	H	B	G	R	C	M	W
T	G	S	R	C	H	B	W	M
W	H	R	M	S	B	G	T	C
C	B	M	W	T	G	S	R	H
B	S	C	H	W	M	R	G	T
H	W	G	T	R	S	M	C	B
M	R	T	G	B	C	W	H	S

So then:

(a) **Samuel Beckett** = C4

(b) **Paris Hilton** = A7

(c) **Diana Ross** = C6

(d) **Paul Simon** = I9

(e) **Desmond Tutu** = D1

(f) **Roy Rogers** = F8

(g) **Emma Thompson** = H4

(h) **Emily Brontë** = F2

(i) **Joseph McCarthy** = G6

(j) **Alec Guinness** = H3

Townsfolk

   (a) Andrew **Lincoln**

   (b) Dionne **Warwick**

   (c) Stuart **Bingham**

   (d) Susannah **York**

   (e) Rod **Hull**

   (f) Don **Cheadle**

   (g) Michael **Bolton**

   (h) Jayne **Mansfield**

   (i) Bing **Crosby**

   (j) Jean **Harlow**

   (k) Harry **Enfield**

   (l) Barbara **Windsor**

 (m) Mick **Fleetwood**

   (n) Thomas **Telford**

   (o) Max **Hastings**

Capital cities

   (a) Irving **Berlin**

   (b) **Brussels** sprout

   (c) **Copenhagen** (horse)

   (d) Dion **Dublin**

   (e) Chicken **Kiev**

   (f) Alex **Kingston**

   (g) **Lima** bean

   (h) Jack **London**

   (i) **Sofia** Vergara

   (j) **Victoria** Wood

   (k) Denzel **Washington**

   (l) Duke of **Wellington**

Fill in the gaps

Lyrics of the song 'Blue Suede Shoes': Wurlitzer 1 4 The Money 2 4
The Show, 3 2 Get Ready **Now** Go **Cat** Go

Sums

a.

$$\text{Bush} - \frac{\text{Ford}}{\text{Hoover} - \text{McKinley} - \text{Madison}} = (41 \text{ or } 43) - \frac{38}{31 - 25 - 4}$$

$$= 22 \text{ or } 24 = \textbf{Cleveland}$$

b.

$$\frac{\text{Moscow} \times \text{Rio de Janeiro} - \text{Barcelona} \times \text{Sydney}}{\text{Munich} - \text{Mexico City}} = \frac{1980 \times 2016 - 1992 \times 2000}{1972 - 1968}$$

$$= 1920 = \textbf{Antwerp}$$

Appendix

Periodic Table of Elements

Key:
Atomic Number
Symbol
Name

1 IA 1A	2 IIA 2A	3 IIIB 3B	4 IVB 4B	5 VB 5B	6 VIB 6B	7 VIIB 7B	8 VIII 8	9 VIII 8	10 VIII 8	11 IB 1B	12 IIB 2B	13 IIIA 3A	14 IVA 4A	15 VA 5A	16 VIA 6A	17 VIIA 7A	18 VIIIA 8A
1 H Hydrogen																	2 He Helium
3 Li Lithium	4 Be Beryllium											5 B Boron	6 C Carbon	7 N Nitrogen	8 O Oxygen	9 F Flourine	10 Ne Neon
11 Na Sodium	12 Mg Magnesium											13 Al Aluminium	14 Si Silicon	15 P Phosphorus	16 S Sulfur	17 Cl Chlorine	18 Ar Argon
19 K Potassium	20 Ca Calcium	21 Sc Scandium	22 Ti Titanium	23 V Varadium	24 Cr Chromium	25 Mn Manganese	26 Fe Iron	27 Co Cobalt	28 Ni Nickel	29 Cu Copper	30 Zn Zinc	31 Ga Gallium	32 Ge Germanium	33 As Arsenic	34 Se Selenium	35 Br Bromine	36 Kr Krypton
37 Rb Rubidium	38 Sr Strontium	39 Y Yttrium	40 Zr Zirconium	41 Nb Niobium	42 Mo Molybdenum	43 Tc Technetium	44 Ru Ruthenium	45 Rh Rhodium	46 Pd Palladium	47 Ag Silver	48 Cd Cadmium	49 In Indium	50 Sn Tin	51 Sb Antimony	52 Te Tellurium	53 I Iodine	54 Xe Xenon
55 Cs Cesium	56 Ba Barium	57-71	72 Hf Hafnium	73 Ta Tantalum	74 W Tungsten	75 Re Rhenium	76 Os Osmium	77 Ir Iridium	78 Pt Platinum	79 Au Gold	80 Hg Mercury	81 Tl Thallium	82 Pb Lead	83 Bi Bismuth	84 Po Polonium	85 At Astatine	86 Rn Radon
87 Fr Francium	88 Ra Radium	89-103	104 Rf Rutherfordium	105 Db Dubnium	106 Sg Seaborgium	107 Bh Bohrium	108 Hs Hassium	109 Mt Meitnerium	110 Ds Darmstadtium	111 Rg Roentgenium	112 Cn Copernicium	113 Nh Nihonium	114 Fl Flerovium	115 Mc Moscovium	116 Lv Livermorium	117 Ts Tennessine	118 Og Oganesson

Lanthanide Series

57 La Lanthanum	58 Ce Cerium	59 Pr Praseodymium	60 Nd Neodymium	61 Pm Promethium	62 Sm Samarium	63 Eu Europium	64 Gd Gadolinium	65 Tb Terbium	66 Dy Dysprosium	67 Ho Holmium	68 Er Erbium	69 Tm Thulium	70 Yb Ytterbium	71 Lu Lutetium

Actinide Series

89 Ac Actinium	90 Th Thorium	91 Pa Protactinium	92 U Uranium	93 Np Neptunium	94 Pu Plutonium	95 Am Americium	96 Cm Curium	97 Bk Berkelium	98 Cf Californium	99 Es Einsteinium	100 Fm Fermium	101 Md Mendelevium	102 No Nobelium	103 Lr Lawrencium

The four elements named Nihonium, Moscovium, Tennessine, Oganesson have not yet been formally approved by the IUPAC Council. This is expected to happen in November 2016.

List of US states	State abbreviations	Capital of the state	Order the state entered the union
Alabama	AL	Montgomery	22
Alaska	AK	Juneau	49
Arizona	AZ	Phoenix	48
Arkansas	AR	Little Rock	25
California	CA	Sacramento	31
Colorado	CO	Denver	38
Connecticut	CT	Hartford	5
Delaware	DE	Dover	1
Florida	FL	Tallahassee	27
Georgia	GA	Atlanta	4
Hawaii	HI	Honolulu	50
Idaho	ID	Boise	43
Illinois	IL	Springfield	21
Indiana	IN	Indianapolis	19
Iowa	IA	Des Moines	29
Kansas	KS	Topeka	34
Kentucky	KY	Frankfort	15
Louisiana	LA	Baton Rouge	18
Maine	ME	Augusta	23
Maryland	MD	Annapolis	7
Massachusetts	MA	Boston	6
Michigan	MI	Lansing	26
Minnesota	MN	Saint Paul	32
Mississippi	MS	Jackson	20
Missouri	MO	Jefferson City	24
Montana	MT	Helena	41
Nebraska	NE	Lincoln	37
Nevada	NV	Carson City	36
New Hampshire	NH	Concord	9
New Jersey	NJ	Trenton	3
New Mexico	NM	Santa Fe	47
New York	NY	Albany	11
North Carolina	NC	Raleigh	12
North Dakota	ND	Bismarck	39

Ohio	OH	Columbus	17
Oklahoma	OK	Oklahoma City	46
Oregon	OR	Salem	33
Pennsylvania	PA	Harrisburg	2
Rhode Island	RI	Providence	13
South Carolina	SC	Columbia	8
South Dakota	SD	Pierre	40
Tennessee	TN	Nashville	16
Texas	TX	Austin	28
Utah	UT	Salt Lake City	45
Vermont	VT	Montpelier	14
Virginia	VA	Richmond	10
Washington	WA	Olympia	42
West Virginia	WV	Charleston	35
Wisconsin	WI	Madison	30
Wyoming	WY	Cheyenne	44

List of US presidents:

1. George Washington	16. Abraham Lincoln	31. Herbert Hoover
2. John Adams	17. Andrew Johnson	32. Franklin D. Roosevelt
3. Thomas Jefferson	18. Ulysses S. Grant	33. Harry S. Truman
4. James Madison	19. Rutherford B. Hayes	34. Dwight D. Eisenhower
5. James Monroe	20. James A. Garfield	35. John F. Kennedy
6. John Quincy Adams	21. Chester A. Arthur	36. Lyndon B. Johnson
7. Andrew Jackson	22. Grover Cleveland	37. Richard Nixon
8. Martin Van Buren	23. Benjamin Harrison	38. Gerald Ford
9. William Henry Harrison	24. Grover Cleveland	39. Jimmy Carter
10. John Tyler	25. William McKinley	40. Ronald Reagan
11. James K. Polk	26. Theodore Roosevelt	41. George Bush
12. Zachary Taylor	27. William Howard Taft	42. Bill Clinton
13. Millard Fillmore	28. Woodrow Wilson	43. George W. Bush
14. Franklin Pierce	29. Warren G. Harding	44. Barack Obama
15. James Buchanan	30. Calvin Coolidge	

NATO phonetic alphabet:

A	–	Alpha/Alfa	N –	November
B	–	Bravo	O –	Oscar
C	–	Charlie	P –	Papa
D	–	Delta	Q –	Quebec
E	–	Echo	R –	Romeo
F	–	Foxtrot	S –	Sierra
G	–	Golf	T –	Tango
H	–	Hotel	U –	Uniform
I	–	India	V –	Victor
J	–	Juliet/Juliett	W –	Whiskey
K	–	Kilo	X –	X-Ray
L	–	Lima	Y –	Yankee
M	–	Mike	Z –	Zulu

Greek alphabet:

Alpha	Nu
Beta	Xi
Gamma	Omicron
Delta	Pi
Epsilon	Rho
Zeta	Sigma
Eta	Tau
Theta	Upsilon
Iota	Phi
Kappa	Chi
Lambda	Psi
Mu	Omega

Morse code, letters and numbers:

A	·−	N	−·	0	−−−−−		
B	−···	O	−−−	1	·−−−−		
C	−·−·	P	·−−·	2	··−−−		
D	−··	Q	−−·−	3	···−−		
E	·	R	·−·	4	····−		
F	··−·	S	···	5	·····		
G	−−·	T	−	6	−····		
H	····	U	··−	7	−−···		
I	··	V	···−	8	−−−··		
J	·−−−	W	·−−	9	−−−−·		
K	−·−	X	−··−				
L	·−··	Y	−·−−				
M	−−	Z	−−··				

Scrabble table of values and frequencies

2 blank tiles	(scoring 0 points)
1 point:	E×12, A×9, I×9, O×8, N×6, R×6, T×6, L×4, S×4, U×4
2 points:	D×4, G×3
3 points:	B×2, C×2, M×2, P×2
4 points:	F×2, H×2, V×2, W×2, Y×2
5 points:	K×1
8 points:	J×1, X×1
10 points:	Q×1, Z×1

Semaphore alphabet:

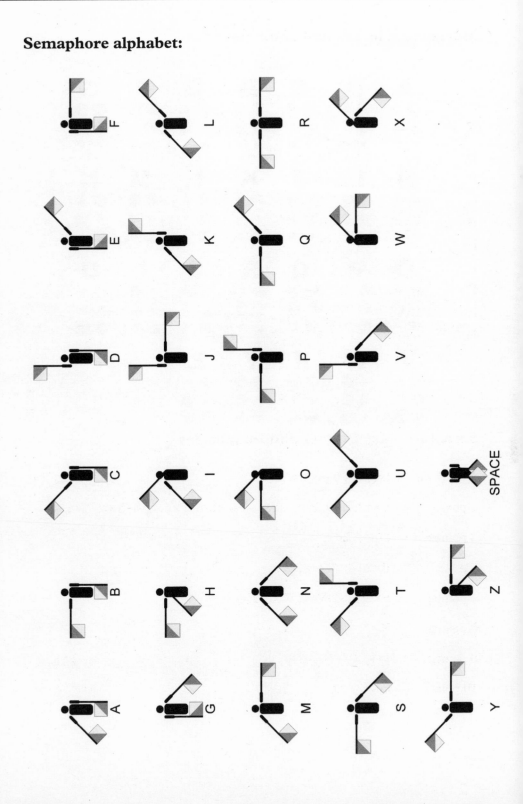

Braille alphabet:

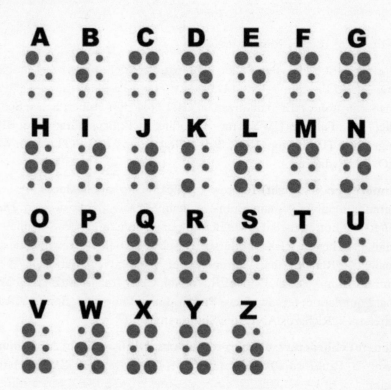

Picture Credits

Puzzle Hunt illustrations: Michael Davis

Alphabetimal: AGX: Ben Salter; IIR: Armin Kübelbeck; CPZ: Kim Hansen; ADO: Adrian Pingstone; CEO: Travelwayoflife; ELR: Jans canon; ORW: Adrian Pingstone; AEF: Von.grzanka; GGL: Joachim Huber; EEM: Scott Wylie; NRT: Falense; UWY: Travelwayoflife; ACF: Brian Gratwicke; BEO: Von.grzanka; HLN: Jjron; KOR: Brian Gratwicke; ADT, EGT: Keith Weller; AMO: Gandydancer

Competition: (left–right) Top row: public domain, public domain, gdcgraphics, public domain, public domain; 2nd row: public domain, *Daily Mail*/REX/Shutterstock, Howie Berlin, public domain; 3rd row: public domain, Patche99z, REX/Shutterstock, public domain; 4th row: public domain, Eva Rinaldi, Allan Warren, Siebbi; 5th row: Neil Grabowsky/ Montclair Film Festival, Derrick Rossignol, Getty Images/Stringer, John Kobal Foundation; 6th row: Steve Punter, Larry Philpott, J. Berliner/BEI/ Shutterstock, Richards/ANL/Rex/Shutterstock

Colourful characters: a) Tourprogolfclubs, b) Alisonnik, c) Arngrímur Jónsson, d) Paul Bird, e) public domain, f) Dan Wooller/REX/Shutterstock, g) gdcgraphics, h) PASOK, i) REX/Shutterstock, j) ANL/REX/Shutterstock, k) REX/Shutterstock, l) Dan Shao, m) public domain, n) Glenn Francis www. pacificprodigital.com, o) Angela George

Next in sequence: 2) Wapcaplet/Angr, 3) Maximilian Dörrbecker

Identify and divide into groups: a) S Meddle/ITV/Rex/Shutterstock, b) Joost Evers/Anefo, c) Dillan Stradlin, d) Disney, e) public domain, f) Colorsport/ REX/Shutterstock, g) Дмитрий Голубович, h) John Mueller, i) Disney, j) Columbia Pictures, k) Steve Back/ANL/REX/Shutterstock, l) Ken McKay/ REX/Shutterstock, m) David Fisher/REX/Shutterstock, n) Rodwey, o) Disney, p) public domain, q) Danielager96, r) public domain, s) Chatham House, t) Eva Rinaldi, u) public domain, v) Howard Lake, w) Daigo Oliva, x) Tim Rooke/ REX/Shutterstock, y) REX/Shutterstock, z) Robin Hume/REX/Shutterstock

Identify the people: a) REX/Shutterstock, b) Walter Bellamy/Stringer, c) Mohan at Doha Stadium Plus, d) public domain, e) Chris Phutully, f) Dan Wooller/REX/Shutterstock, g) Kevin Airs, h) Angela George, i) Fraser Gray/ REX/Shutterstock, j) AVRO, k) public domain, l) public domain, m) Dezo Hoffmann/REX/Shutterstock, n) public domain, o) Gregory Pace/BEI/ Shutterstock, p) public domain, q) Richard Young/REX/Shutterstock, r) public domain, s) Riccardo Ghilardi, t) Ken McKay/ITV/REX/Shutterstock, u) Robert Scoble, v) Glenn Francis, w) public domain, x) Mike Morbeck, y) Kyle Cassidy, z) Panio Gianopoulos

Pictures: (left–right) Top row: public domain, Paypwip, Johann Jaritz, Martijn; 2nd row: Nandhp, jgstore, Sneeboer

Celebrity Sudoku: In the grid: A9) Georges Biard, B2) Chatham House, B7) public domain, C1) Moviestore/REX/Shutterstock, C2) Heinrich Klaffs, C3) Gage Skidmore, D5) Stuart Crawford, D6) public domain, D8) REX/Shutterstock, D9) public domain, E5) Allan Warren, E6) Lou Wolf49, E7) public domain, F4) public domain, F5) David Muscroft/REX/Shutterstock, F7) Damien Everett, G1) Georges Biard, G2) Philippe Berdalle, G4) Clément Bucco-Lechat, G5) Zip2.0, G8) Biser Todorov, H2) Keith Allison, H6) Georges Biard, H8) Brunetto1995, H9) History Archive/REX/Shutterstock, I1) Pace/BEI/BEI/Shutterstock, I4) Roland Godefroy. Below the grid: a) Roger Pic, b) Toglenn, c) public domain, d) Startraks Photo/REX/Shutterstock, e) Kristen Opalinski, f) Orange County Archives, g) Justin Harris, h) public domain, i) public domain, j) Allan Warren

Townsfolk: a) Angela George, b) Allan Warren, c) DerHExer, d) Everett/REX/Shutterstock, e) REX/Shutterstock, f) Bob Bekian, g) Alterna2, h) Kraingkrai Kane, i) public domain, j) public domain, k) Ken McKay/ITV/REX/Shutterstock, l) Matthisvalerie, m) Brandt Luke Zorn, n) public domain, o) *Financial Times*

Capital cities: a) public domain, b) Tiia Monto, c) public domain, d) Ray Tang/REX/Shutterstock, e) Jason Lam, f) Florida Supercon, g) public domain, h) public domain, i) Yahoo, j) Mines Advisory Group, k) Georges Biard, l) public domain

Fill in the gaps: (left–right) public domain, Jeff Belmonte, Thomas Radzak, Jacklee, public domain

Sums: (left–right) A): Abu Shawka, (top) OSX, (bottom) Che, Frank K., johnthescone. B): (top) Ludvig14, Paul Mannix, Bernard Gagnon, Adam.J.W.C, (bottom) Stefan Kühn, Jeff Kramer. Answers: a) Aeroplanepics0112, b) Fuss.

Braille alphabet: DIPF

The styles of Where? and Which? questions are owned by the puzzle setters.

Scrabble, Cluedo, Monopoly and Trivial Pursuit are all trademarks of Hasbro.

Credits

The puzzles in this book were created by the staff of GCHQ in their spare time.

Puzzle editor: Colin

Assistant puzzle editors: Chris, John, Mike, Richard

Principal puzzle setters: Andy, Celia, Chris, Colin, John, Katy, Mike, Nigel, Richard

Additional puzzles set by: Andrew, Andy 2, Daniel, Hannah, Ian, James, James 2, Joanna, Margaret, Nick, Paul, Pete, Tom, Tom 2, Will

Thanks also to: Andy 3, Ian 2, Julian, Ruth, Sam, Sarah

Particular thanks go to Graeme for masterfully co-ordinating it all behind the scenes, and also to Andrew 2, Emily, Faye, Jo, Rebecca, Tony and Wayne for their assistance.

Thanks also to Adam Gauntlett, Jonathan Sissons and the PFD agency, and the team at Penguin: Daniel Bunyard, Fiona Crosby, Sophie Elletson, Catriona Hillerton, Clare Parker and Katie Bowden.

Strenuous efforts have been made by both Penguin and GCHQ to ensure that this book is free from errors. However, it contains a lot of intricate detail and it would be arrogant of us to assert that we have completely succeeded. Any errors we find post publication will be listed on the website www.gchqpuzzlebook.co.uk.

The last seven pages of this book have been left blank for your notes and workings.